Variable Stars

Variable Stars

Michel Petit

with a Foreword by
Paolo Maffei

Translated from the French by
W. J. Duffin

JOHN WILEY & SONS
Chichester · New York · Brisbane · Toronto · Singapore

First published in French as Les Etoiles Variables

Copyright © Masson, Editeur, Paris 1982

English edition Copyright © 1987 by John Wiley & Sons Ltd.

Library of Congress Cataloging in Publication Data:

Petit, Michel, 1935–
 Variable stars.

 Translation of: Les étoiles variables.
 Bibliography: p.
 Includes index.
 1. Stars, Variable. I. Title.
QB835.P4313   1987   523.8'4   86-4060

ISBN 0 471 90920 3

British Library Cataloguing in Publication Data:

Petit, Michel
 Variable stars.
 1. Stars, variable—observers' manuals
 I. Title   II. Les étoiles variables.
 English
 532.8'44   QB835

 ISBN 0 471 90920 3

Printed and bound in the United States of America.

Contents

Translator's Note

For this English edition, the author has provided new material incorporating advances that have been made since the original French edition. In particular, some modifications to the classification and nomenclature of variable stars announced only in mid-1985 have had to be added as an Appendix (E).

Other minor changes have been made to the French text with the approval of the author. These mainly concern the designation of apparent magnitudes: m_{pg} has been used for photographic magnitudes and m_{pv} for photovisual magnitudes instead of p and v, respectively.

The translator is very grateful to Dr G. A. Steigmann of the University of Hull, who has read the entire manuscript and made a number of suggestions for the improvement of the text and, in some cases, for more appropriate terminology.

Foreword

Anybody with an interest in astronomy will be aware of the importance of studying variable stars, a category embracing a wide variety of different types associated with critical stages in stellar evolution.

There is normally little or no change in the brightness of a star. For instance, while we are now sure that the solar constant is subject to rapid rhythmic variations, these amount to no more than 2% of the mean value and do not lead us to classify the sun as a variable star.

As long as a star is located along the main sequence of the H–R diagram, it produces energy by transforming hydrogen into helium: it is in equilibrium and, as required by the Vogt–Russell theorem, it has a well-defined surface temperature and luminosity which depend on its mass and chemical composition. However, when equilibrium conditions come to an end (or, conversely, when they have not yet been reached), the star passes through an unstable phase manifested by variations in its luminosity or its spectrum. Other physical parameters can then be derived, together with information about the structure and aspect of the star at various stages in its life. It is thus possible, from observations of variable stars, to study the formation of stars and the first phase of their life up to the time when they leave the main sequence and their brightness begins to vary, as with the Cepheids or the red giants. They can also be followed in their final stages, when they become, depending on their mass, white dwarfs, neutron stars or perhaps black holes. Variations in white dwarfs and neutron stars have only been discovered in the last few years. As for black holes, the only evidence that they even exist is provided by the observation of their variability and it can only be by a deeper study of this type of observation that such existence could be confirmed or refuted, since it is very unlikely that they could be observed by direct projection on to the celestial sphere because of their very small size. Once again, therefore, it is by the observation of variability that we shall find out whether or not stars can finish their life in such a strange way: swallowed up by a kind of bottomless well!

However, the study of variables is not only important in adding to our knowledge of stellar evolution. As a result of the period–luminosity law, the intrinsic brightness of the Cepheids is known, so that their absolute magnitudes can be deduced from observations of their periods and their real distances can be obtained. If such a star belongs to a cluster or a galaxy, its distance is also known, together with that of all the other objects within it. Moreover, there are several families of Cepheid-type variables: classical, W Virginis and RR Lyrae. The first of these are the youngest and are part of a stellar population formed fairly recently or even still in the process of formation (Population I); the others are older stars belonging to Population II. The various types of Cepheids are therefore useful in giving an indication of population types as well and thus help us to gain a better knowledge, not only of the life of a star but of its very evolution, an essential feature of the evolution of the Universe as a whole.

Variable objects also exist that are not stars. The first of these were discovered several years ago with modest instruments showing them only as points, but recent observations with more powerful and better adapted instruments have revealed them to be the compact and highly luminous nuclei of distant galaxies. Quasi-stellar objects or quasars are the latest of these to be observed and a large number of them are variable. Such variability indicates by its very existence their small relative sizes, even though they are much larger than individual stars.

Briefly, then, the study of variable objects is fundamental in modern astronomy because of its many applications, and it demands that observation should be as continuous as possible since the mode of variation of a star may change with time or become different in nature as the interval between observations becomes longer. For instance, irregular variable stars observed for a long time may reveal some periodicity. Others showing regular pulsation and ignored for some time may then exhibit a significant reduction in their amplitude. Others may have a brightness that increases or diminishes slowly. Yet how can all of them be followed, when currently we know about 30 000 of them with the number increasing every day?

Professional astronomers can afford to spend only a small part of their working life studying such stars. It is true that they are able to make spectroscopic observations or measurements with various radiations outside the atmosphere using artificial satellites or radio telescopes, but they do not always have the telescope time available to follow fluctuations in brightness. This is something that can be done, however, by amateur astronomers with visual or photographic observations made with modest instruments. Many variable stars can be observed with the naked eye or with simple binoculars: although a 10 cm diameter telescope would enable stars of magnitude 11.7 to be seen and one of 20 cm would bring in those of magnitude 13, such a telescope is not easy to obtain and use nowadays. It is not difficult, either, for an amateur to locate the stars to be observed, even faint ones, because celestial maps are available and there are many international organizations that publish the times at which observations should be made and also publish the results obtained. It is a more delicate operation to plan a good programme of observations and to choose the variables that it is important to study. Up-to-date and detailed documentation is needed for such programmes and this is not easy to obtain, even for professional astronomers, because the results of observations and the theories about variables are scattered over numerous texts, particularly in original notes and fundamental articles available only in observatories. These notes and articles are almost useless for a general view of the problem because they exist in such large numbers, and only by carrying out the long and arduous work of sifting through them and gathering together the results can they be put into a useful form.

This is the sort of work that has been carried out by Michel Petit in this book, which is one of the fullest that exists on the subject and currently the most up-to-date. It will prove of enormous help to the amateur wishing to observe variable stars because of the great wealth of statistical information contained within it, together with the detailed characteristics of individual stars.

The book will be of use not only to amateurs, however, but also to anybody with an interest in astronomy because, although variable stars are discussed in all astronomy books, the data presented are not always entirely based on observation. When we read strange and wonderful theoretical conjectures about variable stars, we ought at the same time to look at a text containing observations more than anything else, for only then can we make a more accurate assessment of the ideas and conclusions from such theories.

This book by Michel Petit will enable such assessments to be made, perhaps in a less attractive form than that of a text meant for continuous reading, but in the end more useful. Anybody studying astronomy, even if not explicitly concerned with variable stars, should be grateful to the author and publisher who have produced such a valuable work.

Paolo Maffei
Professor of Astrophysics
University of Perugia, Italy

Introduction

The expression 'variable star' is generally used to describe a star whose *brightness* is not constant. However, this interpretation is rather restrictive, since the fluctuations in the luminosity of a star, known as its 'photometric variation,' are in fact usually accompanied by variations in other physical quantities which may be directly measurable, such as the spectral type, or indirectly deducible, such as the effective temperature.

The study of variable stars as a whole forms one of the major branches of stellar astronomy. It is also one of the most interesting because of the wide variety both of the types of fluctuation encountered and of the problems posed by them. In addition, and this is by no means the least important aspect, it is worth emphasizing that it is one of the rare fields of astronomy in which the amateur with modest instruments can still do useful work. There is a simple reason for this: many variable stars, both those that are aperiodic and those with long periods, need to be studied over years or even decades. The professional astronomer cannot generally do that since large instruments, which are very costly, must be used for original research. Amateurs, on the other hand, are able to undertake such work if they are willing to, either individually or collectively in the context of programmes established by associations for the observation of variable stars; many examples can be cited of long and exhaustive studies of light curves undertaken by amateurs with instruments that are often very modest, such as small telescopes or even simple binoculars.

There are two small volumes in French dealing with variable stars. The first, entitled *Photométrie et Étoiles Variables*, edited by A. Terzan, was published several years ago by the French Astronomical Society. The second, *Observer les Étoiles Variables, Comment? Pourquoi?* was published in 1980 by the French Astronomical Association. Both are very well written; they explain photometric methods and give excellent advice to amateurs wishing to begin this type of observation. However, they describe only briefly the different types of variability and there is no general publication which at the moment provides detailed documentation.

There are not very many works dealing with variable stars in other languages either. Firstly, there is *Variable Stars* by C. Payne-Gaposchkin and S. Gaposchkin, published in 1938. Although this is a fundamental treatise, it should be appreciated that it is completely out of date. Similarly, *The Story of Variable Stars* by L. Campbell and L. Jacchia, first published in 1941, is very well written in a style accessible to the general reader but, in spite of the success it enjoyed in its time, it has never been reissued and is now too old.

Since the 1960s, several books on the subject have been published with varying levels of technical difficulty and ease of reading. Amongst these are *Variable Stars* by J. Glasby, *Veränderliche Sterne* (translated into English under the title *Variable Stars*) by C. Hoffmeister, another book under the same title by W. Strohmeier, and a book on techniques by B. V. Kukarkin, translated from Russian.

There are only two recent books which include accounts of the great progress made in the last few years. Firstly, a small book by L. Rosino, *Le Stelle Variabili*, which is very interesting but in Italian. Then in 1984 there appeared a new edition of C. Hoffmeister's *Veränderliche Sterne*, rewritten by G. Richter and W. Wenzel. Although an excellent book, it is set at a high technical level and is in German.

There are also specialized works each dealing with a particular category of variable star such as novae and eclipsing binaries, but they are generally advanced technical books and not very accessible to amateurs.

In view of all this, it seemed to me that a small book on variable stars would serve several purposes. I have not attempted to describe observational methods since others have done this better than I could, nor do I advise the amateur to undertake one programme of observations rather than another. My aim has rather been to produce a documented book which may be read by any curious spirit, or which can be consulted by any observer—even by those professional astronomers who are normally far removed from the topic but who need at some time or another to collect information on a particular type of variable star.

I have not been content merely to give a simple description of the types of variable star, but have also tried to explain *why* a star varies in such and such a way: in other words, I have attempted to demonstrate the mechanism behind its variation. Such a mechanism is sometimes complex, and it is not always easy to explain it simply.

The study of variable stars is not only an aim in itself, but it can also be used to study, and even solve, many problems which have been posed, and are still posed, to astronomers.

It is also interesting to see how things have changed in the course of time. Thus, several decades ago, astronomers began to use certain groups of variable stars as indicators of distance. The Cepheids are a well known example: since their absolute luminosity was a function of their period of variation, they could be used for estimating the distances of galaxies or of objects within them. Again, many groups of variables have a particular position in the Galaxy: they have therefore been used to study galactic structure and for that it was necessary to refine the classification of variables so that homogeneous stellar types could be identified. For several years, too, much research work has been centred on the fascinating problems connected with stellar evolution. Variables play an important role in this field, since it is known that stars vary in brightness only at definite stages of their evolution.

In a small volume such as this, all these problems can be dealt with only briefly. For all groups of variables, however, I discuss the population type, the distribution in the Galaxy and, in many cases, problems of evolution.

Astronomy, particularly stellar astronomy, has made enormous progress since artificial satellites and space probes have been available to allow work in regions of the spectrum restricted by the terrestrial atmosphere: the far ultraviolet, X-rays and gamma rays. One chapter of this book deals with variable X-ray stars.

The development of radioastronomy and of astronomy at millimetre and sub-millimetre wavelengths and in the infrared has also led to a number of important discoveries. Here, we can indicate the enormous scientific harvest gathered by I.R.A.S. (the infrared astronomical satellite): in a few months, this enabled a huge number of infrared sources to be studied and it also made some fundamental discoveries, one of the most important being that of protoplanetary systems such as that of Vega. Equipment such as this largely justifies the money spent on constructing it and putting it into orbit. We shall return several times to the results obtained.

The book is divided into five parts. The first, entitled 'General Matters,' consists of only two chapters. The first of these has the limited aim of reviewing ideas concerning magnitudes, spectra and the types of galactic population, and of providing an introduction to the terminology used in the rest of the text. The reader does not need a very extensive knowledge of astronomy, but I must strongly emphasize that the concepts reviewed in this chapter are *indispensable* to the full comprehension of the book and that it is therefore essential to make an effort to grasp them: not in fact a very difficult task.*

The second chapter deals with the classification and description of the major classes of variable star and also gives a brief historical review of this branch of stellar astronomy.

With the second part, 'Pulsating Variables,' we come straight to the heart of the subject by first describing the complex mechanism of the pulsations and then dealing in turn with all the types of variable star which exhibit the effect.This is perhaps rather a difficult part to read, but it is extremely important since the pulsating type are by far the most numerous of variable stars.

---

* Some other ideas are not reviewed in the chapter but are also useful, for example those concerning the determination of distances. In this connection, it should be noted that all the distances given are expressed in parsecs, one parsec being equal to 3.26 light-years.

The third part, 'Eruptive and Cataclysmic Variables,' includes all those objects whose variation in luminosity is the result of an eruption or explosion. Some of these, particularly the novae and supernovae, are among the most spectacular of variable stars.

The fourth part, 'Eclipsing Binaries,' deals with close double stars in which the variation results from the occultation of one of the components by the other. We shall see that these are of great significance in astrophysics.

The fifth part concerns peculiar variable stars which do not fall into any of the previous categories; objects which are curious or difficult to classify, but also pulsars and variable X-ray stars. We end with some objects we call 'monsters': the variable nuclei of certain galaxies.

As much information as possible has been assembled in the text. Numerous light curves and explanatory diagrams are included as illustrations and both should prove a great help in understanding the subject.

It gives me pleasure to express my warm thanks to those who have assisted me: Professor Paolo Maffei, who agreed to write the Foreword; M. Jacques Bernad, who has greatly helped me with his advice and opinions; M. Lionel Reisner, who willingly read the initial manuscript; M. Claude Puig, who accepted the thankless task of reading the final manuscript; and finally all those scientific authors and publishers who have given permission for the reproduction of a large number of illustrations. Their names and the appropriate references are given in the captions to the Figures.

Part 1

GENERAL MATTERS

Chapter 1

A Review of
Fundamental Concepts

Before starting on the study of variable stars proper, it is necessary to review briefly some important ideas concerning the system of magnitudes, the principles of photometry and light detectors, the classification of spectra and the population types. These ideas are *absolutely indispensable* to a proper understanding of the book.

MAGNITUDES

Older astronomers ranked the stars visible to the naked eye in six categories, the first corresponding to the brightest and the sixth to the faintest. However, the development of the first large star catalogue in the middle of the last century, the Bonner Durchmusterung, led astronomers to devise a more precise system. N. R. Pogson in 1856 introduced the concept of magnitude in the following way. If E is the illumination produced by the star in a plane perpendicular to the light rays, such as the pupil of the eye, then the magnitude m is defined by the simple equation

$$m = -2.5 \log_{10} E.$$

It follows from this equation that a difference of one magnitude between two stars corresponds to a brightness ratio of 2.51; a difference of 5 magnitudes corresponds to a brightness ratio of $(2.51)^5$, which is about 100. Zero magnitude is assigned to the brightest stars like Vega; the higher the magnitude, the fainter the star, and the faintest known today have magnitudes 23 or 24; they emit several thousand million times less light than does Vega.*

---

* This can be a surprise to the uninitiated! When it is said that there is a difference of 5 magnitudes between two stars (or that the variation in the brightness of one star is of 5 magnitudes), it means that there is a ratio of 100 between the two brightnesses. But for 10 magnitudes, the ratio is $100 \times 100 = 10\,000$; for 15 magnitudes it is 1 million and for 25 magnitudes it is 10 thousand million (10^{10}).

All stars do not have the same colour, so that a red star, for instance, has a photographic magnitude distinctly different from its visual magnitude. We have therefore been led, particularly since photoelectric methods have been used to measure brightness, to define the magnitude for a fixed wavelength. Thus, it was H. Johnson and W. Morgan in 1950 who established a system of magnitudes known as the *UBV* system: the *U* magnitude (near ultraviolet) is measured using appropriate filters at a wavelength of 3600 Ångstroms (Å), *B* (blue) is measured at 4400 Å and *V* (visual) at 5500 Å. Apparent magnitudes are also sometimes denoted by m_U, m_B and m_V, and in line with this there are also the designations m_{pg} and m_{pv} respectively for photographic magnitudes measured with ordinary film, and photovisual magnitudes measured using film that is most sensitive in the same yellow–green spectral region as the eye. It is essential to be aware of this system, which is used throughout the book.

The work of Johnson, Kron and others has enabled other standards to be defined in the red (R, at 7000 Å) and the infrared (I, J, K, L, M and N from 9000 to 35 000 Å). We shall sometimes find examples of these here, but the *U*, *B* and *V* measures are the most common.

A colour index can be defined as the difference in magnitudes between two standards, such as $B-V$ or $U-B$. These have great importance in astrophysics since they describe the energy ratios between two different spectral regions.

In order to compare the *intrinsic brightness* or *luminosity* of two stars, the concept of *absolute magnitude, M*, has been established. This is defined as the magnitude that a star would have if placed at a standard distance from the Earth. This distance has been chosen as 10 parsecs. We point out for the record that the absolute magnitude of the Sun in the Johnson–Morgan system is 4.85 in *V*, 5.50 in *B* and 5.60 in *U*.

Another point to remember is that the *bolometric magnitude* of a star corresponds to the energy radiated over the whole spectrum. To pass from a visual magnitude to a bolometric one (either absolute or apparent) we apply what is called the bolometric correction (*BC*):

$$BC = M_{bol} - M_V.$$

This correction is a function of temperature, and is large for very hot or very cold stars, but zero for stars of spectral type F5. The absolute bolometric magnitude of the Sun is 4.75; in other words, the bolometric correction to be applied to the magnitude in *V* is −0.10:

$$BC_{Sun} = 4.75 - 4.85 = -0.10.$$

METHOD OF OBSERVATION OF VARIABLE STARS

Whether the observations are visual, photographic or photoelectric, a variable star is always studied using the same method as that first described by Argelander: it is compared with one or more stars called comparison stars, and the difference between the magnitudes of the two is determined.

Of course, there are precautions to be taken: the chosen comparison star must have a brightness close to that of the star to be observed, since inaccuracies tend to increase with increasing difference in brightnesses. For convenience, it is also necessary to choose a comparison star not too distant from the variable to be studied. A component of a close binary should be avoided as a comparison star, since its companion could well falsify the measurement. Finally, it is preferable to compare stars whose colour indices are not too different.

It is not always easy to satisfy all these conditions simultaneously, particularly when the variable is located in a region with few stars. Nevertheless, the choice of good comparison stars is important because the accuracy of observations depends largely upon it.

JULIAN DATES AND THE DECIMALIZATION OF THE DAY

It is sometimes difficult to specify the time that elapses between two events, particularly if they are separated by several years or several decades. A system conceived by Joseph Scaliger, known as Julian dating, is therefore used. Days are counted from an origin as zero, chosen to be as long ago as 1st January of the year 4713 B.C. Thus, 1st January 1980 is the Julian date (abbreviated to JD) 2 444 239. Note that the Julian day begins at 12 noon UT and not at 12 midnight.

Extensive conversion tables exist, and Appendix B gives a month-by-month table of Julian dates from 1975 to 1999.

In addition, to specify the times taken for variations, it is convenient to decimalize the day, and a table of decimal values for every 10 min of the day is provided in Appendix C. It is easy to extrapolate should a greater precision than this be needed. To take an example: an observation made at 21 h 50 min UT on 1st October 1980 would be expressed (since the Julian day starts at 12 noon) by JD 2 444 514.40972.

LIGHT DETECTORS

The first of these detectors is the human eye and until the last century astronomical observations were entirely visual.

The eye operates over an extensive range of light intensity: there is a ratio of approximately 1000 million between the maximum and minimum light fluxes that it can record. Not only that, but it is sensitive to very small differences in brightness of the order of a tenth of a magnitude. In favourable environments, its angular resolution (the angle subtended at it by the smallest discernible object) is about 3 minutes of arc, which corresponds to an aircraft with a 10 m wingspan flying at an altitude of 10 000 m.

The second detector is the photographic plate. This is distinctly less sensitive than the eye, but it has a considerable advantage over it: it accumulates the light received. This is what enables objects too faint for the eye to see to be photographed by using exposure times that are sometimes considerable. The photographic plate remains unequalled in programmes for systematic observation of the sky.

Electronic devices form the most recently developed types of detector. Several kinds are used, but rather than give a detailed description of them we shall limit ourselves to an outline of the main systems.

First of all, there is the photocell or photomultiplier. The photons of light strike a layer of a conveniently chosen element and detach electrons from it. These create an electric current whose strength is proportional to the amount of light flux and which is very greatly amplified by a suitable device. Many types of cell are produced commercially, and they are in general use in observatories and are becoming more and more so by amateurs.

An interesting instrument is the electronic camera, developed at the Paris Observatory by A. Lallemand. Electrons detached from a caesium plate are focused by electron lenses and an image is then formed on a special photographic emulsion. This instrument performs extremely well; it enables exposure times to be used that are 100 times shorter than those with classical telescope photography. However, it is difficult to use, firstly because it requires vigorous cooling with liquid nitrogen and secondly because the tube containing the photocathode must be maintained at a very high vacuum. This explains why photomultipliers are preferred for photometry: their performance is inferior, but they are cheaper and more convenient to use.

'Photon counters' appeared in the 1970s. These are ultra-sensitive and can detect only a few photons; the amplification between the input and output flux is of the order of 10. A device forms an optical image on a vidicon tube; in other words, the observer sees the image gradually being formed on the screen as the photons arrive at the counter. This equipment has one defect, although only a minor one: it cannot be used to observe very bright objects since it becomes completely saturated.

A great improvement is now being achieved with the CCD (charge-coupled device) system. The principle is as follows: the apparatus contains a series of electrodes which trap the electrons liberated by the impact of photons on a silicon plate. After the exposure, electronic processing allows the electrons which have accumulated in the traps to be made to 'slide' as far as a reading head. This then transmits the information to the memory of a computer.

These devices, highly miniaturized, can be coupled to a photon counter, thus avoiding the use of a vidicon tube and producing a considerable improvement in the sensitivity of the system. For very weak fluxes, the performance is improved by a factor of 100, so that there is a gain of about 5 magnitudes. Not only that, but such instruments can operate at very short wavelengths (X-rays and gamma rays). In the future, they will be used to equip space vehicles, particularly the Space Telescope of 2.40 m aperture which must be able to record objects of magnitudes 28 or 29.

SPECTRAL CLASSIFICATION

It was at the beginning of the century that E. C. Pickering and his collaborators established the spectral classification which is still used today, known as the Harvard classification:

$$W - O - B - A - F - G - K - M - S - C$$

The system is summarized in Table 1, from which it can be seen that there is a decrease in temperature from W stars to C stars. (Note that class C has replaced two types, R and N, which were formerly used.)

Table 1. The Harvard spectral classification

Type	Characteristics	Temperature (K)	Examples
W	Ionized helium (He II) in emission spectra. Carbon (WC) and nitrogen (WN) present	50 000	V444 Cyg (WN5) CV Ser (WC7)
O	Neutral helium (He I) or ionized helium in absorption spectra. Some hydrogen lines	35/20 000	ς Pup (O5) δ Ori (O9.5)
B	Helium weaker; hydrogen (Balmer series) strong	20/12 000	ε Ori (B0) β Ori (B8)
A	Balmer lines dominant: many metallic lines particularly ionized calcium (Ca II)	12/8000	α Lyr (A0) α Aql (A7)
F	Neutral calcium (Ca I) and ionized calcium. Other metallic lines.	8/7000	α Per (F5) β Vir (F9)
G	Ca II dominant. Many metallic lines	7/5000	Sun (G7) β Aql (G8)
K	Ca II and metallic lines. Molecular bands appear	5000/3500	σ Dra (K0) α Tau (K5)
M	Titanium oxide (TiO) dominant. Other molecular bands	3500/3000	α Sco (M1.5) α Her (M5)
S	Zirconium oxide (ZrO) dominant. Otherwise resembles M	3000	R Gem (S3) R And (S7)
C	Intense bands of carbide molecules (CH, CN)	2500	19 Psc (C6)

Since the number of stars that are observed spectroscopically is so enormous (more than 500 000 in all), it has long been the practice to create sub-classes such as B0, B1, B2 up to B9, for example. More recently, intermediate subdivisions have been introduced such as B0, B0.5, B1, B1.5, ..., and even in some cases finer subdivisions such as B0, B0.25, B0.5, etc.

The Harvard classification does not distinguish dwarfs from giants and a lot of work has been done in attempting to specify the different luminosities of stars that may be of the same spectral type. The classification most commonly used today is that of Yerkes, known as the MK system after its originators Morgan, Keenan and Kellman.

Figure 1 shows what is called a Hertzsprung–Russell (H–R) diagram. In such a diagram, the absolute magnitude M_V is plotted along the vertical axis and the spectral type along the horizontal axis. The lines drawn in the figure represent the distribution of stars in the MK system.

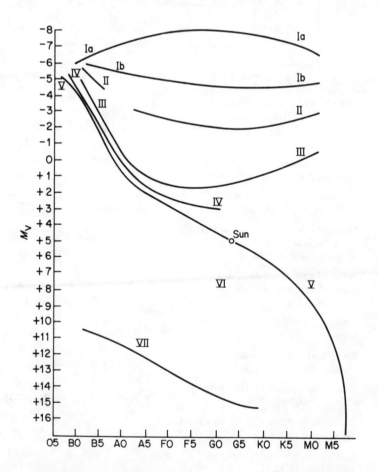

Figure 1.   Absolute magnitude as a function of spectral type: MK system

The Roman numerals indicate the *luminosity classes.* Ia and Ib correspond to two very distinct groups of supergiants, II to bright giants and III to normal giants. Class IV corresponds to subgiants while class V is the main sequence (dwarfs) which includes the great majority of stars and extends from O stars to red dwarfs: an older notation that is still sometimes used denotes dwarfs by the letter d before the spectral type (e.g. dM4 instead of M4 V). Class VI is restricted to stars called subdwarfs, an ill-defined group whose most distinct characteristic is a spectrum poorer in metallic lines than that of a main sequence star. Finally, class VII is that of the white dwarfs or degenerate stars.

It will be seen that W stars do not appear in the diagram. These stars are sub-luminous and are located on the extreme left below the main sequence. As for C and S types, they seem to occur mainly in classes II and III.

Many examples using the MK classification will be found in this book, and it is therefore essential to become familiar with it. It is also useful to be aware of a few symbols that indicate peculiarities in spectra. The letter e indicates the presence of emission lines in the spectrum (we shall find, for example, the notations A5 Iae, B4 Ve or M3 IIIe), and the letter p indicates a non-classified peculiarity (e.g. A0 p).

In addition, the letter Q is used to denote the spectra of novae while q indicates that a star has a 'nebular' spectrum, similar to that of novae in certain stages of their evolution. Class VII is not very often used to denote white dwarfs and the letter D is generally used instead. Thus, DA corresponds to a white dwarf whose spectrum has an energy distribution similar to that of an A star.

STELLAR POPULATIONS

The concept of stellar populations was introduced by W. Baade in 1944. It gives an indication of the location of stars in the Galaxy and also of their age. Baade considered two main types:

Population I: These are stars in the galactic arms; they are young or even very young, since some are still being formed. They are always very close to the mean galactic plane. In other words, Population I forms a flat system.

Population II: These include the stars in the central region of the Galaxy, those of the galactic halo and in the globular clusters. These stars may be located at a considerable distance from the galactic plane and they are said to form a spherical system. They are 'old' stars, the first to be formed in our Galaxy.

In fact, this description of the classification is slightly perfunctory. Nowadays, a very young Population I, called the Extreme Population I, is contrasted with one that is less young, called the Intermediate Population I. It is the same with Population II.

It is not our intention here to develop any further the concept of population types, which depends on differences in chemical composition between the various groups, but it is necessary to be aware of it at least in outline, since it is a concept associated with the notion of age and it recurs frequently in this book.

Chapter 2

General Points about Variable Stars

Following the review of fundamental concepts that was the subject of the first chapter, we now reach the topic that is our main concern: the study of variable stars. In this chapter, we shall give a brief historical review of the discovery of variables, and then go on to describe both the system of notation used to designate them as well as the large catalogues of variable stars that exist. However, we must first of all deal with their classification, which must be kept in mind if the phenomena described in later chapters are to be understood.

CLASSIFICATION OF VARIABLE STARS

This problem has occupied the attention of astronomers for a long time, since it is essential for advancing our ability to understand the observed phenomena.

We shall not discuss the earliest classifications, which were all based on the forms of the variations alone. It was two American astronomers, S. and C. Payne-Gaposchkin, who in 1938 proposed a classification based both on the apparent fluctuations in brightness and on the physical characteristics (spectra and absolute magnitude) of the stars.

It is true that recent discoveries have led to modifications in certain classes and to the introduction of new ones, but in broad outline the classification remains valid today.

PULSATING VARIABLES

Here, the fluctuations in luminosity are caused by a pulsation of the star. This is due to the combination of two forces: radiation pressure, which tends to expand the star, and gravitation, which does the opposite. When the star is at its maximum diameter, radiation pressure is at its minimum, the temperature is low and the gaseous velocity is zero. Gravity then makes the star contract.

The pressure then increases, and so do the temperature and the gaseous velocity. At its minimum diameter, the pressure will make the star 'inflate:' in other words, the temperature and the pressure will then diminish and when the maximum diameter is reached, the whole process starts again. The repetition period is often very regular. We shall be able to return to a consideration of pulsation phenomena, which affect very different types of stars, in Part 2.

Table 2. The principle types of variable star (see also Appendix E)

Type	Symbol	Period	Amplitude	Remarks
Pulsating variables				
RR Lyrae	RR	0.2–1 d	0.5–1.5	Spectra : A or F
Cepheids	C	1–50 d	0.2–2.0	Sp: F or G. Several types
RV Tauri	RV	30–150 d	1.0–3.0	Sp: G or K
Semi-regular	SR	20–500 d	0.2–2.5	Sp: M, C, S. Several types
Long period	M	100–600 d	2.5–10	Sp: M, C or S
Irregular	L	Non-periodic	0.2–2.0	Sp: M, C or S
β Canis Majoris	βC	0.1–0.5 d	<0.2	Sp: B0 to B4
Dwarf Cepheids	RR	0.05–0.25 d	0.5–1	Sp: A or F
δ Scuti	δ Sc	0.05–0.2 d	<0.2	Sp: A or F
α Canum Venaticorum	α CV	0.5–>100 d	<0.2	Sp: peculiar A
ZZ Ceti	ZZ	1–15 min	<0.05	White dwarfs
Eruptive variables				
Novae	N		7–>15	Explosion of part of stellar atmosphere
Supernovae	SN		>20	Stellar explosion
Novoids	N1			Several types
Dwarf novae	UG,Z	10–500 d	2–6	Repeated explosions
Nebular variables	In,Is		1–4	Interaction with gaseous nebulae
	UVn		1–4	Flare stars
Red dwarfs	UV,BY		0.1–5	Sp: K or M. Flare stars
Eclipsing binaries				
Algol variables	EA	0.1–10000 d	0.1–3	All sp. present
β Lyrae	EB	0.5–200 d	0.1–1.5	Giants O, B and A
W Ursae Majoris	EW	0.2–1 d	0.2–1	Dwarfs F, G or K

ERUPTIVE (OR CATACLYSMIC) VARIABLES

The brightness here increases abruptly following an eruption which affects all or part of the star's atmosphere. This increase is accompanied by a rise in temperature and hence by major variations in the spectra. When the maximum brightness has passed, the star wanes and slowly cools.

ECLIPSING BINARIES (OR GEOMETRICAL VARIABLES)

Here, it is no longer a question of a physical variation but of occultation: one of the components of a close pair of stars passes in front of the other, thus eclipsing it and reducing the total brightness from the binary system.

Each of these three great classes includes stars that differ among themselves in the form and amplitude of the luminous variations as well as in their spectra and absolute magnitudes. The principal types of variability distinguished in this way are given in Table 2, together with the abbreviations used for them. In general, these types form the headings and subheadings of the chapters which follow (see Appendix E for recent modifications in designations).

The existence of a good scheme of classification is essential to the understanding of the phenomena observed. For that reason, the classification is modified or supplemented from time to time. Thus, the Variable Stars Commission of the International Astronomical Union decided in 1979 to get rid of three types or sub-types which did not correspond to physical reality. On the other hand, they have added new types when their existence has been confirmed.

THE DISCOVERY OF VARIABLE STARS

The history of variable stars is intimately linked with that of astronomy in general. The oldest astronomical observations that have come down to us, those from China, give an account of the sudden appearance of extremely bright stars. Some of these were novae or supernovae, but a study of each case has shown that it was often comets that were seen.

A very bright star appeared in Scorpio in July of 134 B.C. If Pliny the historian is to be believed, it was this that made Hipparchus of Rhodes decide to undertake his star catalogue—yet it seems certain that this was not a star but a comet!

Some of these 'new stars' have been extremely bright, like that of July 1054 which appeared in Taurus or that of November 1572 in Cassiopeia. It is now known that these were supernovae.

On 13 August 1596, David Fabricius discovered in the constellation Cetus (the Whale) the first *periodic* variable, named by Helvetius in 1638 as 'Mira' (the Wonderful). Montanari revealed the variability of Algol in 1669 and its period was determined by Goodricke in 1782. However, it seems very likely that the fluctuating brightness of this star had already been noticed in the Middle Ages by Arabic astronomers: indeed, its name comes from the Arabic for 'The Demon.'

Until about 1850, discoveries were quite rare. Apart from novae and supernovae, only 13 variable stars were known at the beginning of the nineteenth century. These are listed in Table 3, with the date of their discovery and the names of those who first observed them.

The publication from 1855 onwards of Argelander's great stellar catalogue

Table 3. The first known variable stars

Stars	Date	Observer
o Ceti	1596	Fabricius
β Persei	1669	Montanari
χ Cygni	1687	Kirch
R Hydrae	1704	Maraldi
R Delphini	1751	Hencke
R Leonis	1782	Koch
μ Cephei	1782	Herschel
β Lyrae	1784	Goodricke
δ Cephei	1784	Goodricke
η Aquilae	1784	Pigott
R Coronae Borealis	1795	Pigott
R Scuti	1795	Pigott
α Herculis	1795	Herschel

(Bonner Durchmusterung), containing the positions and magnitudes of more than 300 000 stars, led to the discovery of a certain number of variables. In addition, the catalogue established by Chandler in 1896 already contained 393 variables, almost all discovered visually.

The introduction of systematic searches for variable stars using photography at the end of the last century considerably quickened the rate of discovery. It rapidly became necessary to include in catalogues only those stars whose variability had been confirmed and to give a definitive designation only to variables in the Galaxy.

In spite of these precautions, the number of catalogued variable stars grew considerably, as is evident from the following figures for five of the catalogues published in this century: 1916, *Geschichte und Literatur* (Müller and Hartwig), 1687 variables; 1937, *Katalog und Ephemeriden* (Prager), 6968 variables; 1948, *General Catalogue* (Kukarkin), 1st edition, 10 912 variables; 1958, *General Catalogue*, 2nd edition, 14 708 variables; 1970, *General Catalogue*, 3rd edition, 20 448 variables. With supplements and lists of definitive designations published by the International Astronomical Union, a figure of 28 457 catalogued variables was reached by 1981. The accelerating rate of discoveries is well illustrated by Figure 2, even though it is limited to catalogued variables.

In fact, a still greater number of variables are known. So-called Suspected Variables should also be included, i.e. those whose variability is often slightly doubtful and which are not yet catalogued because additional observations are needed. The *New Catalogue of Suspected Variable Stars* contains 14 210 of them but, because the catalogue stopped in 1980, there are many others. To these must be added variables associated with globular clusters (more than 2500), those discovered in the Magellanic Clouds (more than 3500) and those (several thousands) that have been found in nearby galaxies. In all, there must be in excess of 50 000 known variables.

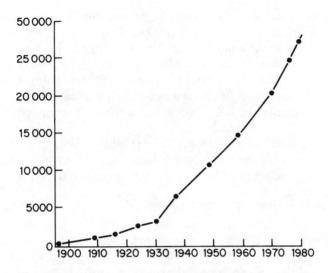

Figure 2.   Growth in the number of catalogued variable stars

RATIFICATION OF VARIABILITY AND SYSTEM OF DESIGNATION

Variables do not receive their definitive designation until the variability is confirmed and the type of variability is known. The designation is given by the Variable Star Commission of the International Astronomical Union,[*] which publishes every year a list of new definitive designations.

The system of designation used goes back to Argelander. In order to distinguish variable stars from others, he denoted the first variable discovered in a constellation by R, the second was denoted by S, the third by T and so on up to Z. The variables known at that time were certainly not very numerous!

When the nine letters R to Z were used up, which for some constellations occurred around 1880, the letters were doubled to give RR, RS, RT, ... , RZ, then SS, ST, ... , SZ, TT, TU, ... , TZ, up to the fifty-fourth variable, ZZ.

With discoveries accelerating rapidly, it was decided to designate those after the 54th as follows: AA, AB, AC, AD, AE, ... , up to AZ, BB, BC, BD, BE, ... , up to BZ, CC, CD, CE, ... , up to CZ, and so on up to QZ, omitting J.

This method brought the number of available designations up to 334 for each of the 88 constellations. This number, apparently considerable, was nevertheless found to be insufficient for the rich constellations of the Milky Way or those situated in the direction of the Galactic centre. The method then adopted is one proposed by Charles André at the end of the last century and consists simply of calling them V1, V2, V3, etc., as they are discovered.

---

[*] The IAU has more than 40 Commissions, and the Variable Star Commission is No. 27.

From the 335th variable, therefore, the designation is by the letter V followed by a number: V335, V336, V337, etc. This system is unlimited and it is a pity that it was not adopted from the very beginning.

In the constellation richest in variables, Sagittarius, we have now exceeded V4000! In Ophiuchus, the number has passed V2000, in Cygnus, Aquila and Orion, V1000; and V335 has been reached in 15 or so others. In contrast, the poorest constellation is Caelum in the southern hemisphere, where the last variable discovered was Z.

It should be noted that, when they are variables, the bright stars denoted by Greek letters using the Bayer system have not received other designations. They occur in modern catalogues with their original letters: for example, α Scorpii, β Persei, δ Cephei, etc.

Newly discovered variables that have not yet received their definitive designation receive a provisional one from the observer who discovers them, while awaiting final ratification. This complicates the task of those studying variable stars since one star, discovered at several observatories, can receive two or three different provisional designations at the same time.

CATALOGUES AND DOCUMENTATION

A large amount of work has been done on variable stars, but it is of very uneven quality. Any observer wishing to carry out useful research must have as much information as possible about the variable to be studied and catalogues form one source of such information.

Before the last war, the Germans published a catalogue called *Katalog und Ephemeriden der veränderlichen Sterne*. It contained all the variables that had received a definitive designation together with their principal characteristics. First edited by R. Prager, it was continued by Schneller and its last edition was in 1943.

In addition, Müller and Hartwig published from 1916 to 1922 a work entitled *Geschichte und Literatur der veränderlichen Sterne* containing a complete bibliography of the 1687 variables catalogued in 1916. This was reissued under the title *Geschichte und Literatur des Lichwechsels der veränderlichen Sterne* by Prager, and later by Schneller, and published in several volumes between 1934 and 1963.

After the Second World War, the publication of a general catalogue was left to Soviet astronomers. So, in 1948, there appeared the *General Catalogue of Variable Stars* (abbreviated to G.C.V.S.), published in Russian and English under the direction of B. V. Kukarkin and P. P. Parenago. The third edition was published from 1969 onwards, with supplements appearing until 1976 (6 volumes in all). The fourth edition, prepared by P. N. Kholopov and his collaborators, is currently in the course of publication. It contains very complete information on the 28 457 variables catalogued in 1981, as well as several thousands of references and supplementary lists for the old novae, the extragalactic supernovae, pulsars, quasars, variable galaxies and so on.

B. V. Kukarkin and his collaborators also published a *Catalogue of Suspected Variable Stars* (abbreviated to C.S.V.) in both 1952 and 1965, which contained more than 12 000 variables not yet catalogued. These volumes have been replaced by the *New Catalogue of Suspected Variable Stars* (N.S.V.), published in 1982 under the direction of P. N. Kholopov, which contains information about 14 810 stars awaiting ratification.

In view of the interest in purely bibliographical works such as those of Müller and Hartwig and of Prager and Schneller, two astronomers from the Sonneberg Observatory, H. Huth and W. Wentzel, have prepared a new work called the *Bibliographic Catalogue of Variable Stars*, which contains an enormous amount of information and a large number of references. This was published in 1981 by the Centre for Stellar Documentation at Strasbourg Observatory.

Observers of variable stars thus have considerable documentation at their disposal. These catalogues, particularly the General Catalogue, are working tools that are absolutely indispensable to anybody, whether amateur or professional, who wishes to carry out useful research.

Part 2

PULSATING VARIABLES

The class of so-called 'pulsating' variables, which is numerically the largest, includes groups of stars which are very different from each other but whose variation in brightness stems from the same process, their alternate expansion and contraction.

Some astrophysicists have used the name 'instability strip' for a region of the Hertzsprung–Russell (H–R) diagram which groups together the white or yellow pulsating variables of spectral types A, F or G; in other words, the dwarf Cepheids and the δ Scuti stars, the RR Lyrae stars, the classical Cepheids and the W Virginis stars.

Another branch of the H–R diagram consists of orange and red variables with longer periods, and includes the RV Tauri stars, the semi-regular or irregular red stars, and the long-period Mira type variables. All these can be seen in the H–R diagram in Figure 3.

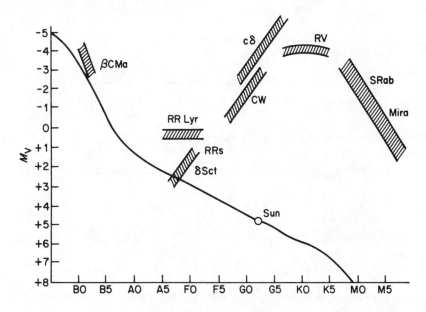

Figure 3.   Location of the various groups of pulsating variables in the H–R diagram

Pulsation was described briefly in Chapter 2 of Part 1 and we return to it in more detail below. Following that are five chapters on the various types of pulsating variables, the first chapter dealing with RR Lyrae stars and the second with the various types of Cepheids. The third describes the long-period variables; the fourth, the groups of semi-regular and irregular variables; and the fifth, certain groups of pulsating variables that are numerically small but are sometimes important because of the interesting problems they pose that are well worth studying.

THE MECHANISM OF PULSATION

We have already said a little about this process in connection with the classification of variables. A star is subject to two opposing forces: radiation pressure which tries to expand it, and gravitation which tries to collapse it. In normal stars these two balance each other. Under certain equilibrium conditions, however, the star alternately expands and contracts, the two forces taking turn in becoming predominant. We shall take a precise example: that of δ Cephei, a Cepheid variable, which will serve as a model for all the other pulsating stars. Its properties are illustrated in Figure 4.

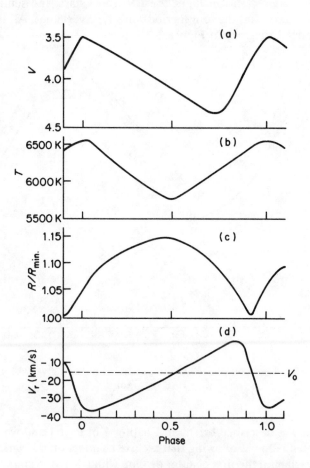

Figure 4. The pulsation of δ Cephei. (a) The light curve; (b) variation of temperature (K); (c) variation of the radius R; (d) variation of radial velocity. At the value V_0 in (d), the star has reached its maximum or minimum radius which is therefore instantaneously invariant. All data are given for one complete period from one maximum to the next, that is, for one cycle

First of all, consider the time at which the star has its maximum radius (1.14 times the minimum value), which is about 18 million kilometres or, if preferred, 26 solar radii. At this instant, the temperature is nearly at its minimum (about 5800 K), the diameter is not varying at all and the radial velocity is represented by the value V_0 in Figure 4d. The same applies to the minimum points.

The star begins to contract slowly at first and then more quickly, at the same time becoming hotter and brighter. When the radius reaches its minimum, the temperature and brightness are nearly at their maxima. The radial velocity, which was positive during the period of contraction, now reverses as the new expansion phase begins. Just after this, the brightness and temperature pass through maxima and then begin to decrease. There is a small time lag between each of the phenomena.

To fix ideas, take the case of δ Cephei. The amplitude of its variations in brightness is not very large: 0.86 magnitude (in V). This means that the ratio of its maximum to its minimum brightness is about 2. The effective temperature varies from 6600 K at the maximum to 5800 K at the minimum, and this produces a change in spectral type from F5 Ib to G1 Ib. The overall amplitude of the variations in radial velocity is about 40 km/s and the radius itself oscillates between 16 and 18 million km (between 23 and 26 solar radii).

It is not our intention here to go into the theory of these pulsations, a theory established in 1919 by A. Eddington and developed later by many others; it is complex and involves a knowledge of the internal structure of the stars. There is, however, one thing that should be pointed out because it explains something that we shall meet several times in the following chapters.

Physically, a pulsation is a form of oscillation which originates at an internal point of the star and is carried to its surface. The oscillation can be produced in what is called the fundamental mode or in one or more harmonics, depending on the position of the point of origin with respect to the surface. To take the simple analogy of an organ pipe: a vibration can be produced with a particular fundamental wavelength (and thus a particular pitch) depending on the length of the pipe. Any given pipe may, however, also produce harmonics at higher frequencies and shorter wavelengths if suitably excited.

The oscillation produced in the star must be maintained or it will die away. Theory shows that if it is to be maintained the star must have a sort of 'reservoir' whose function is to store the heat when the star is hot and contracted, so as to be able to restore it progressively as the star cools and expands.

Astrophysicists have shown that it is mainly the internal layers of ionized helium that play the role of a reservoir. Why is that? Because during the contraction phase, the liberated energy is used in such a layer to ionize atoms and not to heat the gas. This layer remains cooler than the non-ionized layers surrounding it; it can thus absorb more heat. During expansion, the heat is restored and this supplies additional energy for maintaining the pulsation.

In the same star, there can be several ionized layers located at different depths. If this is the case, it can have several pulsation periods which overlap, and this produces a complex light curve. We shall see shortly that this is a situation that often occurs in all types of pulsating variable.

Chapter 1

RR Lyrae Stars

RR Lyrae variables exist in large numbers. They were once known as cluster variables because the first of them were observed in globular clusters, but the General Catalogue currently lists about 6000 throughout the Galaxy.

The typical star is RR Lyrae, the brightest of this group. It varies from a visual magnitude of 7.06 to one of 8.12 with a period of 13 h 36 min or 0.56682326 day. The light curve, shown in Figure 5, is highly asymmetric,

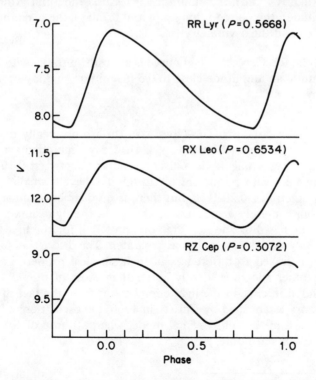

Figure 5. Light curves for three prototype stars: RR Lyr, RX Leo (RRab) and RZ Cep (RRc)

with the rise to the maximum taking 2 h 35 min and the fall taking about 11 h. The variation in luminosity, as with all pulsating stars, is accompanied by a significant variation in colour: the colour index $B - V$ changes from $+0.14$ at the maximum to $+0.45$ at the minimum. It follows from this that the amplitude of 1.06 in visual magnitude corresponds to one of 1.37 in the B magnitude. The large variations in spectra should also be noted, with the spectral type varying from A8 to F7.

THE SUB-TYPES

Astronomers (notably Solon Bailey in 1915) have for a long time separated the RR Lyrae stars into the following three sub-types whose light curves are illustrated in Figure 5:

  (i) RRa (like RR Lyr), characterized by an amplitude near to, or greater than, 1 magnitude, by a period of about 0.5 day and by marked asymmetry.

 (ii) RRb (like RX Leo): the amplitude is less than the first group (0.5–0.8 magnitude), the period is longer (about 0.7 day), the maximum is more rounded and the asymmetry less marked.

(iii) RRc (like RZ Cep): the light curve is almost sinusoidal with a rounded maximum, an amplitude close to 0.5 magnitude, and a period of about 0.3 day.

It will be seen later that sub-types a and b are not really distinct. It is true that the variables with periods of 0.4–0.5 day generally have an amplitude greater than 1 magnitude and an asymmetry between 0.10 and 0.20, whereas the stars with periods of 0.7 to 0.8 day have a smaller amplitude and an asymmetry of 0.20–0.25, but there is no distinct boundary between the two groups: merely a progressive change in the light curve. The General Catalogue has thus adopted the notation RRab for the first two Bailey sub-types, only lists the RRc class as separate.* The distinction between the two groups is linked to differences in their physical properties: RRc stars are slightly bluer and thus slightly hotter than those of the RRab class. In addition, the difference in the light curves stems from the fact that, according to a theory established by Christy in 1966, the oscillations of the RRab stars take place in the fundamental mode, whereas those of the RRc stars occur in the first harmonic.

---

* The General Catalogue also has an RRs sub-type. We shall see in Chapter 5 that these stars, known as dwarf Cepheids, are distinct from the RR Lyrae stars.

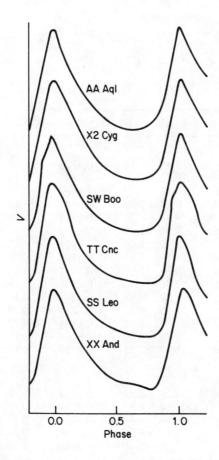

Figure 6. Light curves for six RRab variables: AA Aql ($P = 0.361$ day), XZ Cyg (0.466 day), SW Boo (0.513 day), TT Cnc (0.563 day), SS Leo (0.629 day) and XX And (0.722 day). The curve changes as the period varies

Figure 6 shows shows the light curves for six RRab stars and Table 4 gives the properties of 15 RRab and 5 RRc stars. The asymmetry D is the ratio of the time occupied by the rising part of the curve to that for the complete cycle. V_{max} is the maximum visual magnitude and ΔV the amplitude of the fluctuations.

Table 4. 20 RR Lyrae star

Designation	V_{max}	ΔV	Period (days)	D	Spectra
RRab					
RS Boo	9.69	1.13	0.3773	0.20	A7–F5
TW Her	10.52	1.29	0.3996	0.13	A7–F5
VZ Her	10.72	1.32	0.4402	0.12	A6–F4
SW And	9.14	0.95	0.4422	0.17	A7–F8
S Ara	9.96	1.24	0.4518	0.16	A3–F3
RR Leo	9.96	1.34	0.4523	0.13	A7–F5
XZ Cyg	9.00	1.16	0.4664	0.17	A6–F6
AR Her	10.59	1.04	0.4700	0.20	A7–F3
TW Boo	10.64	1.04	0.5322	0.13	F0–F8
RR Lyr	7.06	1.06	0.5668	0.19	A8–F7
RX Eri	9.22	0.88	0.5872	0.16	A7–F6
RV Cet	10.35	0.87	0.6234	0.25	F0–G5
RX Leo	11.56	0.71	0.6534	0.23	F2–F7
SU Dra	9.25	0.99	0.6604	0.18	F0–F7
XX And	10.08	1.05	0.7227	0.19	A8–F6
RRc					
DH Peg	9.27	0.51	0.2554	0.38	A1–A7
SX UMa	10.62	0.54	0.3071	0.38	A4–F5
RZ Cep	9.06	0.68	0.3086	0.32	A0–A9
T Sex	9.82	0.50	0.3246	0.42	A9–F4
BV Aqr	10.80	0.40	0.3640	0.40	A4–A9

DISTRIBUTION OF PERIODS

Figure 7 is a histogram of the periods of 4057 RRab and 319 RRc stars compiled by the author.

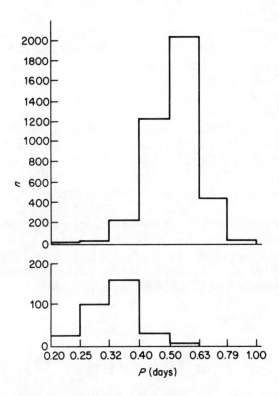

Figure 7. Histograms showing the distribution of periods for 4057 RRab stars (upper diagram) and 319 RRc stars (lower diagram)

For the RRab stars, the shortest known periods are about 0.22 day and the maximum in the frequency distribution (the modal value of the period) is around 0.55 day. This maximum is very pronounced with half the periods being between 0.50 and 0.63 day. The longest periods exceed 1 day (XX Vir: $P = 1.3482$ day), but there are very few of these and they are not given in the histogram. It is difficult to define a boundary between the RR Lyrae stars and the Cepheids; indeed, we shall see in the next chapter that some stars classed as RR Lyrae by their light curves can be considered as Cepheids when the distribution of energy in their spectra is examined.

For the RRc stars, the periods are shorter and the modal value occurs around 0.34 day, with extreme values at 0.20 and 0.55 day. It should be pointed out that these stars are much less numerous in the Galaxy than those of the RRab sub-type.

VARIATIONS IN PERIOD

Many RR Lyrae stars show variations in their periods which may be continuous or may occur abruptly. Take the case of RZ Cep, of sub-type RRc, which was one of the first stars studied from this point of view. From 1890 to 1901, the period was 0.308692 day or 7 h 24 min 31.03 s. In August 1901, it decreased by 5.86 s (to 0.308625 day) and remained at that value for 15 years. In November 1916, it increased by 3.88 s and then in November 1923 by 1.81 s. The period was then 7 h 24 min 30.86 s and it stayed at that value until 1948. There was then another change, and then in 1952 the period became 7 h 24 min 29.50 s. In June 1963, it was no more than 7 h 24 min 26.90 s (0.308645 day).

It might be thought that a change which affected only the fifth or sixth decimal place in the fraction of the day would be difficult to detect. Nothing of the sort: in fact, RZ Cep has 1184 maxima in a year and a change of 1 s in one period would produce a total change of 20 min per year, and this is certainly detectable. Over 10 years, such a change would amount to over 3 h.

We shall see in the next chapter how the variations in period are measured by taking a particular case, but it can already be seen that changes in the period, although very small, can be detected by a process of accumulation.

A large number of RR Lyrae stars exhibit variations in their periods and over a hundred cases have been studied. Such variations are always small and seem to be cyclic, with cyclic periods in some cases amounting to thousands or tens of thousands of days.

THE BLAZHKO EFFECT

We have seen that some stars have several pulsation periods. This produces a continuous deformation of the light curve, called the Blazhko effect after the Soviet astronomer who first studied the problem.

A classic example is that of AR Her, examined by J. Balasz and L. Détré.* Figure 8 shows the deformations in the light curve very clearly. The amplitude varies by 0.4 magnitude and the asymmetry also varies considerably, changing from 0.12 to 0.31. Not only that, but these variations are cyclic and a modulation period $P_b = 31.5$ days has been calculated (that is, 67 times P_1), after which the variable star is back in its original state.

Some cases of the Blazhko effect have been studied very thoroughly and, to mention only a few, we may quote the following: RR Lyr, whose modulation period is 40.812 days); BV Aqr (11.25 days); RW Cnc (29.9 days); SW And (36.83 days); AT And (82.75 days); and AR Ser (110.7 days). The modulation period is generally between 50 and 200 times the primary period.

---

* The periods determined by these workers were $P_1 = 0.47002$ day and $P_2 = 0.46311$ day.

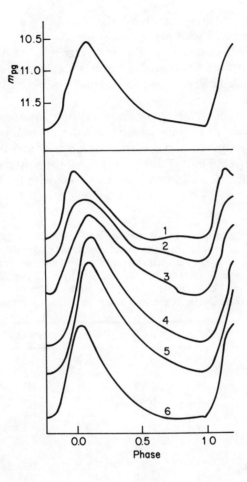

Figure 8.   The Blazhko effect in AR Herculis. Upper diagram: average light curve. Lower diagram: six different cycles (from J. Balazs and L. Détré, *Mitteilungen der Sternwarte, Budapest*, No. 8, 1939)

The Blazhko effect occurs among all types of pulsating variable, but it affects mainly those of short period.

RR LYRAE STARS IN GLOBULAR CLUSTERS

Many RR Lyrae stars occur in globular clusters and it is this which gave them their original name of cluster variables.* There are about 130 globular clusters in the Galaxy, some of them with no RR Lyrae stars, others among

---

* Globular clusters also contain other types of variable star, particularly Cepheids, which are dealt with in the next chapter.

the larger and brighter ones such as NGC 104 (47 Tucanae) or NGC 6205 (Messier 13, Herculis) with just a few. Some, however, like NGC 5139 (ω Cen) or NGC 5272 (Messier 3), contain a large number of them.

A great deal of work has been done in this area and we cannot examine it all in detail. It has, however, yielded certain facts and relationships that are worth mentioning.

It is possible to obtain the spectrum of a cluster considered as a single star and this gives what is called the 'mean spectrum.' Now there exists a distinct correlation between the mean spectrum of a cluster and the number of RR Lyrae stars discovered in it. Table 5 shows that the clusters of spectral types F0 to F8 are much richer in variables than those with the later-type spectra (G0 to G6).

Table 5. Galactic location, mean spectra and number of variables for various globular clusters. R is the distance from the cluster to the Sun expressed in kiloparsecs (kpc), b is the galactic latitude and Z is the mean distance of the cluster from the mean galactic plane, also in kpc. Z can be calculated from the simple equation $Z = R \sin b$. It is positive if the object is to the north of the galactic plane and negative if it is to the south

Globular cluster	R (kpc)	b (°)	Z (kpc)	Mean sp.	No. of variables
NGC 4590	9.6	+37	+5.8	A7	42
NGC 7078 (M15)	9.8	−27	−4.4	F2	119
NGC 7006	39.4	−19	−12.8	F3	74
NGC 5024 (M53)	16.5	+80	+16.3	F4	47
IC 4499	17.5	−21	−6.3	F5	169
NGC 5904 (M5)	7.1	+47	+5.2	F6	97
NGC 5139 (ω Cen)	5.2	+15	+1.3	F7	179
NGC 5272 (M3)	9.6	+78	+9.3	F7	214
NGC 6715 (M54)	21.9	−14	−5.3	F7	80
NGC 6266 (M62)	6.5	+7	+0.8	F8	74
NGC 6402 (M14)	14.5	+14	+3.3	G1	40
NGC 6626	14.9	−6	−0.5	G1	3
NGC 104 (47Tuc)	4.3	−45	−3.0	G3	2
NGC 6838 (M71)	3.6	−5	−0.1	G6	4

It can be seen from Table 5 that there is a relationship between b (or Z) and the number of variables: clusters far from the galactic plane have in general an abundance of RR Lyrae stars, whereas those near the plane have relatively few.

A large number of stars are subject to variations in their period and to the Blazhko effect. Several reports, the first of which was due to Sterne (1934), have demonstrated the existence of a strict correlation between the density of the cluster and the frequency of the variations in period. In addition, several workers including Szeidl have shown that in the same cluster there are some periods which are increasing and others decreasing, often cyclically.

The distances of these globular clusters have been determined by various methods and are well known. The absolute magnitudes of the RR Lyrae stars within them can therefore also be determined and are found to vary from 0.5 to 0.9 in *V*. RR Lyrae stars in the rest of the Galaxy whose distances can be obtained have values of the same order, 0.6 on average. Note that there is no appreciable difference in absolute magnitude between sub-types ab and c.

It is important to realize that, unlike other pulsating stars, particularly the Cepheids, there is no relationship between the period and the absolute magnitude for RR Lyrae stars. Some workers, such as Dickens in 1970, have shown that the absolute magnitude is connected rather with the concentration of metallic elements in the star's atmosphere, those with the lowest intrinsic brightness being the richest in metals (particularly ionized calcium Ca II).

Table 6 shows some properties of ten globular clusters chosen among those which are richest in RR Lyrae stars. The total number of variables is repeated from Table 5, and in addition the number of RRab and RRc stars that have been closely studied is given, together with their mean periods. The last column gives the number of RRc stars as a percentage of the total studied.

Table 6.   Distribution and mean periods of RR Lyrae stars in globular clusters

Globular cluster	No. of RR Lyrae stars	RRab		RRc		RRc (%)
		n	*P*	*n*	*P*	
Group 1						
NGC 7006	72	55	0.566	4	0.294	2.1
IC 4499	169	87	0.558	2	0.302	2.2
NGC 5904 (M5)	99	68	0.546	24	0.319	26.1
NGC 5272 (M3)	214	150	0.553	26	0.328	11.0
NGC 6402	77	37	0.558	3	0.331	7.5
NGC 6266 (M62)	89	62	0.544	3	0.299	4.6
Group 2						
NGC 7078 (M15)	119	69	0.645	36	0.321	34.3
NGC 5024 (M53)	55	24	0.640	20	0.364	45.4
NGC 4590	42	14	0.625	23	0.378	62.2
NGC 5139 (ω Cen)	179	86	0.649	60	0.372	41.1

The distribution of the periods is unusual (Figure 9). In 1939, Oosterhoff showed that two groups of clusters can be distinguished: one in which the mean period of RRab stars is 0.55 day and that of RRc stars is 0.32 day; and another in which the mean periods are 0.64 and 0.36 day, respectively.

It is interesting to note that clusters such as NGC 5024, 5139 and 7078, where the periods of the RRab group are quite close to 0.64 day, are also

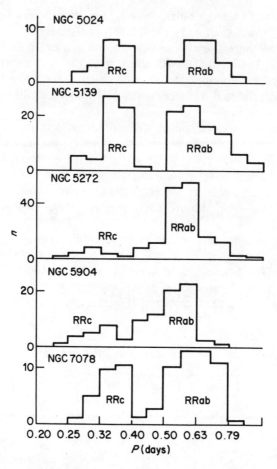

Figure 9.   Distribution of periods for RR Lyrae stars in five globular clusters

those which have the highest percentages of RRc stars, about 40%. There is thus a relation between the mean period of the sub-type RRab stars and the percentage of RRc stars in the cluster.

We end this description of RR Lyrae stars in globular clusters by pointing out that variations in the periods are common. Out of 112 stars examined in Messier 3, Szeidl found 22 whose periods are slightly increasing and 25 with decreasing periods. In Messier 5, Coutts observed 20 periods increasing, 12 decreasing and only 18 remaining constant. In certain quite rare cases, the period decreases and increases cyclically.

DISTRIBUTION IN THE GALAXY

RR Lyrae stars belong to Population II; in other words, there are a great number in the galactic nucleus, in the halo and, as we have seen, in globular clusters. These are the 'old' and less massive stars. It is estimated that their masses are less than half the solar mass (from 0.43–0.48 of the solar mass, according to L. Rosino), whereas the radii are 4–5 times greater than that of the Sun. Their density is therefore very low. A lot of work has been done on the distribution of these stars in the Galaxy. We cannot examine any of it in detail, and give here only a brief summary of the results obtained.

If the Sun is assumed to be 9.5 kpc from the centre of the Galaxy and in the immediate neighbourhood of the galactic plane, the spatial density of RR Lyrae stars can be calculated in stars per cubic kpc. A curve is obtained like that in Figure 10, from which it can be seen that the density decreases slowly away from the galactic plane; there are still RR Lyrae stars at 10 kpc from it. It can be deduced from this that they form a quasi-spherical system. On the other hand, the density increases sharply towards the centre of the Galaxy. At 4 kpc from the centre, the density is four times greater than it is near the Sun. Interstellar absorption prevents us having more accurate results closer in than this, but it is obvious that the density of these stars must increase further as the centre is approached.

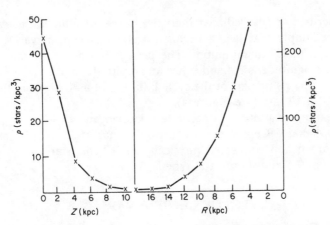

Figure 10. Distribution of RR Lyrae stars in the Galaxy. On the left: the density as a function of distance Z from the galactic plane. On the right: the density in the galactic plane as a function of the distance R from the galactic centre

Variations of the period with position in the Galaxy have also been looked for but, unlike the Cepheids, there is clearly no such variation. It is the same with a possible relationship between the period and the distance from the galactic plane; in other words, the period seems to be completely unaffected by galactic position.

Chapter 2

The Cepheids

The typical star, δ Cephei, has been known since antiquity; it features in the catalogue of Hipparchus compiled in 126 BC. Its variability was detected by Goodricke in 1784 and since then it has been the subject of numerous studies.

We described its detailed characteristics in the introduction to Part 2. We recall here simply that it is a supergiant whose spectrum varies from F5 at maximum to G2 at minimum. The period is 5 d 8 h 47 min 30 s; the maximum magnitude is 3.4 and it has an amplitude in V of 0.83 (1.30 in B, 1.66 in U). The rising part of the cycle lasts 1 d 14 h 30 min and the decline lasts 3 d 18 h 17 min (see Figure 4).

The absolute magnitude is accurately known and is −3.4. It should be pointed out that δ Cephei has a companion star at an angular separation of 41″ from it and connected physically to it. This star has an apparent magnitude of 7.5 and its spectral type is A0 V.

TYPES OF CEPHEID VARIABLE

About 700 Cepheids are known in the Galaxy, but they all fall into two, or perhaps three, large groups. First there are the classical Cepheids, whose typical star is δ Cephei, and which are stars of Population I; in the General Catalogue they are denoted by Cδ. Then there are the W Virginis type, which include stars belonging to Population II; these are denoted by CW.

Several years ago, Efremov discovered a special group of the Population I stars. They are Cepheids with a small amplitude and a sinusoidal light curve. For these, the notation Cδs was proposed, but so far this has not been officially adopted.

CEPHEIDS OF POPULATION I

We examine first the classical Cδ Cepheids. A histogram showing the distribution of the periods of 427 of these stars (Figure 11a) exhibits one very pronounced maximum between 4 and 6 days and another, weaker, one between 12 and 15 days, with a minimum at 9 days. The shortest periods are about 2 days and the longest more than 50 days.

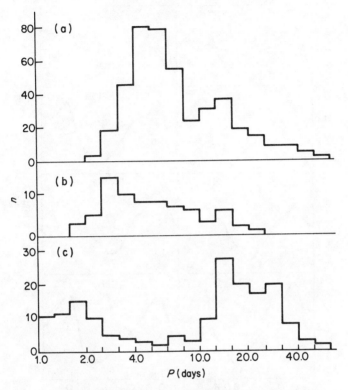

Figure 11.   Histograms showing the distribution of periods: in (a), for 423 Cδ Cepheids; (b), for 73 Cδs Cepheids; (c), for 179 CW stars

Cδ stars thus fall into two groups, separated by the 9 day minimum period. In the first group, the amplitudes of the variations lie between 0.7 and 1.2 magnitude and the asymmetry is 0.25–0.30. For those with shorter periods in this group, the light curves are smooth, but those with cycles of 7–9 days have a distinct 'bump' in the descending part of the curve (Figure 12).

For stars with 10–12 day periods, the amplitude is smaller and the asymmetry approximately 0.4. This gives the light curves a different appearance (Figure 12). For the shorter periods, the amplitude increases and the asymmetry becomes well marked. Around a period of 15 days, the 'bump' appears, this time at the beginning of the rise to the maximum. It disappears towards 25–30 days but occasionally a small nodule is to be seen at the minimum. This change in the light curves as the period varies was noted by Hertzsprung around 1926 and has been studied several times since then, particularly by Kukarkin and Parenago and by C. Payne-Gaposchkin.

38

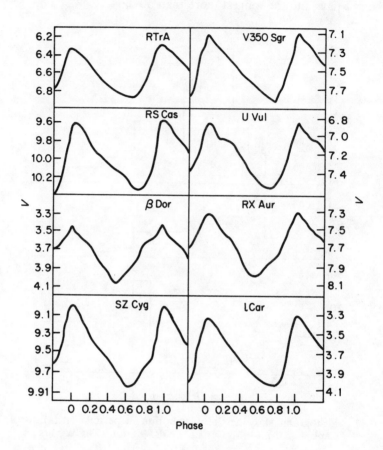

Figure 12. Light curves for eight Cδ Cepheids: R TrA (*P* = 3.3893 days), V350 Sgr (5.1542 days), RS Cas (6.2957 days), U Vul (7.9907 days), β Dor (9.8420 days), RX Aur (11.6235 days), SZ Cyg (15.1099 days) and 1 Car (35.5225 days)

Table 7.   Some bright Cδ-type Cepheids

Designation	V_{max}	ΔV	P (days)	D	Spectra
R TrA	6.35	0.68	3.3893	0.36	F6–G4
RT Aur	5.06	0.79	3.7279	0.25	F4–G1 Ib
T Vul	5.43	0.63	4.4356	0.31	F5–G0 Ib
δ Cep	3.44	0.83	5.3663	0.30	F5–G2 Ib
Y Sgr	5.33	0.76	5.7733	0.25	F6–G4 Ib
R Cru	6.42	0.84	5.8257	0.28	F6–G6 Ib
U Sgr	6.33	0.78	6.7449	0.27	F6–G1 Ib
X Sgr	4.19	0.67	7.0122	0.22	F5–G5 Ib
U Aql	6.06	0.82	7.0239	0.27	F5–G0 I–II
η Aql	3.50	0.87	7.1766	0.29	F5–G2 Ib
W Sgr	4.25	0.81	7.5947	0.22	F5–G4 Ib
W Gem	6.60	0.73	7.9141	0.34	F5–G0 Ib
S Sge	5.19	0.81	8.3821	0.25	F6–G5 Ib
β Dor	3.38	0.70	9.8420	0.49	F6–G2 Ia
TT Aql	6.51	1.12	13.7546	0.34	F6–G5 Ib
X Cyg	5.81	1.20	16.3866	0.35	F6–G5 Ib
T Mon	5.63	1.03	27.0205	0.29	F7–K1 Ia
l Car	3.32	0.79	35.5225	0.38	F6–K0 Ib
U Car	5.82	1.12	38.7681	0.20	F6–K0 Ib
SV Vul	6.73	1.03	45.035	0.20	F7–G9 Ia

Table 7 gives details of 20 well known bright Cδ Cepheids, V_{max} being the maximum magnitude, ΔV the amplitude and D the asymmetry.

The Cδs group, as we have said, is characterized by an amplitude of less than 0.5 magnitude and a not very pronounced asymmetry. The properties of ten of these stars are given in Table 8 and the light curves of four of them in Figure 13. An interesting point is that some of these are spectroscopic binaries and there may therefore be a connection between the small amplitude and the spectroscopic doubling.

The periods (Figure 11b) normally range from 1.8 to more than 20 days. However, it has been shown in the last few years that some can have very small amplitudes, sometimes less than 0.1 magnitude, and very long periods. This is the case with two recently studied variables (1978). The first of these is V925 Sco, observed by A. van Genderen and P. S. The; its period, still not accurately measured, lies between 70 and 80 days and its amplitude is 0.12 magnitude. Secondly, there is V810 Cen, observed by W. Eichendorf and B. Reipurth and belonging to an open cluster (Stock 14). Its period, if confirmed, is the longest one known in the Galaxy, 125 days, and its amplitude is no more than 0.2 magnitude.

Another case is that of R Puppis, which belongs to the open cluster NGC 2439. Its variability was announced in 1879 by B. Gould, then denied, so that for a long time it was considered to be constant. Observations by M. Seift carried out during 1978 showed that it is a Cepheid of very long period with a V amplitude of only 0.06 magnitude.

40

Table 8.   Some bright Cδs-type Cepheids

Designation	V_{max}	ΔV	P (days)	D	Spectra
SU Cas	5.74	0.40	1.9493	0.40	F7–G5 Ib–II
DT Cyg	5.56	0.81	2.4991	0.50	F5–F8 II
SZ Tau	6.38	0.48	3.1487	0.45	F6–F9 Ib
α UMi	1.94	0.11	3.9698	0.50	F7–F9 Ib
AH Vel	5.50	0.36	4.2272	0.49	F8–G2 Ib
FF Aql	5.19	0.34	4.4710	0.45	F5–F8 Ia
AX Cir	5.64	0.42	5.2735	0.40	F5–G2
S Mus	5.90	0.54	9.6601	0.53	F8–G4 Ib
ς Gem	3.68	0.48	10.1508	0.50	F7–G5 Ib
Y Oph	5.93	0.49	17.1241	0.40	F5–G0 Ib

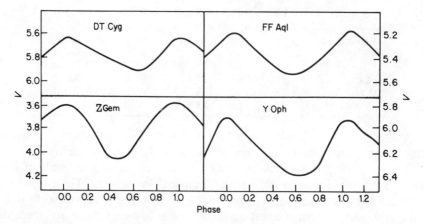

Figure 13.   Light curves for four Cδs Cepheids: DT Cyg (P = 2.4991 days), FF
Aql (4.4710 days), ς Gem (10.1308 days), and Y Oph (17.1241 days)

These last three Cepheids have common characteristics: they are very
bright, much more so than classical Cepheids (absolute magnitude −7 to
−7.5) and are associated with galactic or open clusters. They are therefore
very young.

CEPHEIDS OF POPULATION II

The typical star in this category, W Virginis, was discovered by Schonfeld in
1866. It was long classified along with the other Cepheids, even though the
shape of its light curve differed significantly from those of Cδ Cepheids with
the same period (Figure 14) and in spite of the presence of intense ionized
calcium emission lines in its spectrum at certain parts of the cycle. It also had
the unusual galactic position of +61°, whereas most of the typical Cepheids
were very close to the galactic plane.

The discovery in 1939 of a star resembling W Virginis and belonging to the globular cluster NGC 5139 (ω Cen) showed that a new group of Cepheids had to be created for those distributed in globular clusters and in the galactic halo. At present, nearly 200 galactic variables of the W Virginis type are known. Table 9 gives the properties of a dozen of the better observed of these.

It can be seen from the histogram in Figure 11c that here again there are two groups, depending on their periods:

(i) Those of very short period, less than 2.5 days. The light curves of these stars are sometimes peculiar, like that of VZ Aql (Figure 15), which shows two 'bumps,' one on the rising part of the curve and the other on the decline. The amplitude is often greater than one magnitude and the asymmetry is pronounced.

(ii) Those with periods longer than 10 days, with a maximum occurrence of periods between 15 and 20 days. The amplitude often exceeds one magnitude. For periods up to 20–23 days, the asymmetry is less noticeable and there are 'bumps' on the descending part of the curves which disappear for the longer periods. This can be seen in Figure 15.

It is quite rare to find stars in this category with periods between 2.5 and 10 days.

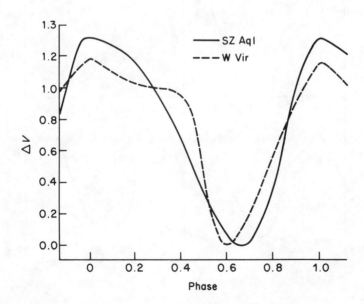

Figure 14. Comparative light curves for SZ Aql (P = 17.138 days) belonging to population I and W Vir (17.274 days) belonging to population II

Table 9.   Some CW-type Cepheids

Designation	V_{max}	δV	P (days)	D	Spectra
SW Tau	9.22	0.83	1.5836	0.25	F4–F8
VZ Aql	13.21	0.83	1.6687	0.25	
TU Cas	7.09	1.05	2.1393	0.26	F3–F5 II
ST Tau	7.79	0.80	4.0342	0.30	F5–G4
κ Pav	3.93	0.89	9.088	0.44	F5–G5
AP Her	10.39	0.82	10.408	0.40	F2–G0 Ib
V410 Sgr	12.10	0.97	13.7733	0.41	F6–G2
W Vir	9.51	1.20	17.2736	0.47	F0–G0 Ib
RU Cam	8.29	1.05	22.354	0.40	K0–R2
RX Lib	11.65	0.79	24.9459	0.39	F9–K0
TW Cap	9.96	1.34	28.6574	0.28	A5–F4 Ib
RR Mic	11.01	0.85	31.7817	0.24	G0

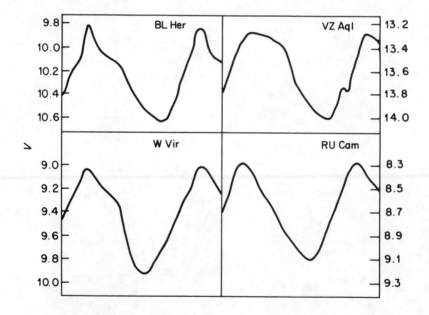

Figure 15.   Light curves for four CW Cepheids, two of very short period (BL Her, 1.3075 days, and VZ Aql, 1.6682 days) and two with long periods (W Vir, 17.274 days, and RU Cam, 22.159 days)

DISTINCTION BETWEEN Cδ AND CW CEPHEIDS

It is not always easy to distinguish these two large groups. There are undoubtedly differences between the shapes of the light curves, the spectra
and in the curves showing variations in radial velocity, but M. Petit showed
in 1960 that the differences are sometimes difficult to bring out. As regards
the galactic position, this does not provide a sufficient criterion since some
stars that are undoubtedly CW type are close to the galactic plane.

An interesting method of distinguishing the two types was put forward
in 1969–70 by Kheilo, Kolesnik and their colleagues. Its principle is very
simple: let ΔV, ΔB and ΔU be the total amplitude of the variations in V, B
and U, respectively. The following ratios or gradients are calculated:

$$G_V = \Delta V/\Delta B \text{ and } G_U = \Delta U/\Delta B.$$

These ratios in fact define the energy distribution in the spectra very simply,
and they have values characteristic of a given type of variable star.

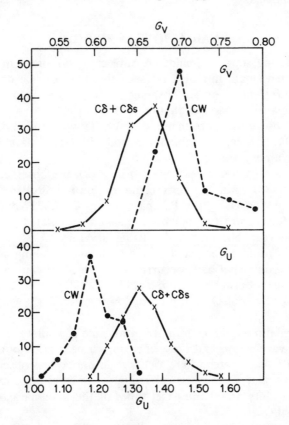

Figure 16.   Distribution of the ratios G_U and G_V for the two
Cepheid populations

Such ratios were calculated by Kolesnik and Kheilo for the stars whose U, B and V curves were known at that time: 106 Cepheids and 116 RR Lyrae. The accumulation of photoelectric observations has enabled me to up-date this work and to give the results covering 379 Cepheids and 183 RR Lyrae stars:

	Mean G_V	Mean G_U
304 Cδ and Cδs	0.657	1.337
75 CW	0.708	1.188
183 RR	0.767	0.980

The dispersion is still large and it can be seen from Figure 16 that the histograms overlap. Nevertheless, this method seems to be one of the best for determining the category of a Cepheid. It can also be used for stars with periods of about 1 day, where some may be RR Lyrae stars and others CW-type Cepheids.

VARIATIONS IN PERIOD

Like all pulsating stars, Cepheids are subject to variations in their period. These are often small, but we have seen that they can be detected by the cumulative difference between observed and calculated values.

Let us take a concrete case, that of T Mon, a Cepheid having a period of 27 days and known for more than a century. Between Julian dates 2415 223 and 2425 194, Parenago obtained a period of 27.0212 days, but this later proved to be a poor value.

A table is constructed (see Table 10) in which the observed maxima are recorded, giving the number zero to the maximum of Julian date 2415 223 (columns 1 and 2). Using the Parenago value of 27.0212 days for the period, a time for each of the maxima can also be calculated (column 3). The difference between observed and calculated values, O−C, is then worked out (column 4) and a corrected period is obtained, taking account of the number of maxima that have occurred (column 5). It can be seen that the real period varies between 27.0053 and 27.0228 days, a total ΔP of 0.0175 days. The variations of O−C are plotted in Figure 17, showing the changes clearly.

In practice, a much larger number of maxima is available for T Mon and this enables verification and cross-checking to be carried out. In Table 10 are given only those dates where a change in period was observed.

This type of analysis must be carried out with care: there must be a high degree of certainty about the exact times of the maxima and this has not always been the case.

The changes in period may be continuous, with a regular increase or decrease, or they may be abrupt, as is the case with DT Cyg. In some cases,

Table 10.   Variations in the period of T Mon (from M. Petit)

Observed max. JD 24+	Serial No. of maximum	Calculated max. JD 24+	O−C (days)	Corrected period (days)
05 230.43	−370	05 225.73	+4.70	
				27.0084
15 223.57	0	15 223.57	0.00	
				27.0212
25 194.38	+369	25 194.39	−0.01	
				27.0053
27 057.75	+438	27 058.86	−1.11	
				27.0206
32 894.21	+654	32 895.44	−1.23	
				27.0228
41 514.51	+973	41 515.20	−0.69	

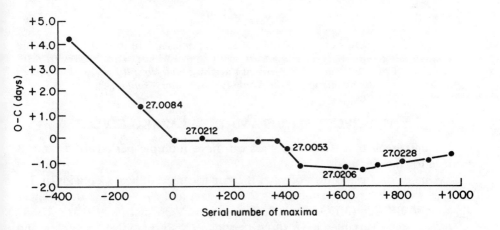

Figure 17.   Variations in the period of T Mon

the period alternately increases and decreases, as it does with T Mon, SZ Cas and IU Cyg. Finally, the variations may be complex: κ Pav has had 12 large changes of period over 80 years, both increases and decreases, the extreme values being 9.0832 and 9.1178 days.

The range of the variations ΔP clearly depends on the total length of time over which the observations have been taken. To make comparisons, a common unit must therefore be used. If we call E the number of maxima, the mean value of the variation ΔP in one day can be written $\Delta P/PE$. There is a clear relationship between this quantity and the period P, as is shown in Figure 18, in which $\Delta P/PE$ is plotted on a logarithmic scale.

The long-period Cepheids often have sizeable variations in their periods, particularly those belonging to Population II where the variation can be as much as 3% of the period (that is, 0.5 day for a period of 17 days).

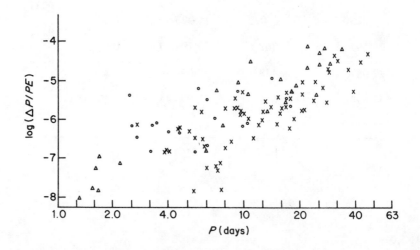

Figure 18. Variations in the periods of 123 Cepheids as a function of P itself. The points plotted with $\times$ refer to Cδ stars, with $\bigcirc$ to Cδs stars and those with $\triangle$ to CW stars. (From M. Petit, Inf. Bulletin of Variable Stars, No. 1402, 1978. *Reproduced by permission of the Konkoly Observatory, Budapest*)

MULTIPLE PERIODS AND THE BLAZHKO EFFECT

We have seen that RR Lyrae stars can have multiple periods; the same is true for the Cepheids.

One of the best known of what the Americans call 'beat Cepheids' is TU Cas, which has two periods $P_0 = 2.1393$ days and $P_1 = 1.5183$ days. There are also BK Cen (3.1739 and 2.2366 days) and V 367 Sct (6.2930 and 4.3849 days). Some variables have three periods: VX Pup studied by Stobie and Ballona, for instance, has $P_0 = 3.011172$ days, $P_1 = 2.1349$ days and $P_2 = 1.8706$ days. It is interesting that the ratio P_0/P_1 is almost constant, in all cases being approximately 0.7. This tends to suggest that all the stars have the same internal structure even if their sizes, and hence their periods, are different.

The Blazhko effect can sometimes be significant. The most spectacular case is that of RU Cam, whose period, otherwise slightly variable, is about 22.159 days. In 1961–62, J. Smak found a significant amplitude of 1.6 magnitude in V. In 1964, he observed that it had fallen to 0.55 magnitude. It continued to decrease and in August 1966 it was no more than 0.08 magnitude, without any appreciable change in the period. It then progressively increased once more, to return to its normal value. On the basis of previous observations, L. Détré has concluded that the phenomenon is periodic with a modulation period of 84 primary periods, that is, slightly over 5 years.

There are many examples of the Blazhko effect, mainly in short-period stars, but only a few of them have been studied up to the present time.

THE PERIOD–LUMINOSITY RELATIONSHIP

Important discoveries are often made by chance. At the Harvard Observatory in 1912, H. Leavitt had just determined the magnitudes and periods of 25 Cepheids that she had discovered in the Small Magellanic Cloud. Before preparing a table for publication, she chose to grade these Cepheids in order of increasing period, and noticed that they were then also in order of increasing brightness. Since the Small Magellanic Cloud is a single galaxy and a great distance away from ours, and since it is also very small in size compared with its distance away, it could be assumed that all the stars in that galaxy are at approximately the same distance from us. There seemed therefore to be a relationship between the period and the brightness, and the famous period–luminosity relationship had been discovered.

It was necessary, however, to verify that this relation also applied to galactic Cepheids. The problem is not simple, since the Cepheids are very distant supergiants and none lends itself to distance measurement by trigonometrical parallax. H. Shapley chose eleven bright Cepheids whose proper motions could be determined. He deduced from them the absolute magnitudes used as a basis for calibrating the period–luminosity relationship. It has been possible ever since to use the Cepheids to determine the distances of any galaxies in which they can be observed.

Unfortunately, the position of the zero point of the Shapley relationship was not accurate, since nobody knew at the time that there were two main types of population and two populations of Cepheids!

In 1952, W. Baade undertook a study of the RR Lyrae stars in the Andromeda galaxy M31 using the Mount Palomar telescope. To his great surprise, he found them 1.5 magnitudes fainter than they should have been! The reason for that is simple: the distance of M31 had been determined using Cepheids, but the zero point of the Shapley relation was incorrect so that the distance obtained was too small. It was necessary to multiply by 2, or even slightly more. This had a rather odd consequence: the nearby galaxies in which Cepheids could be observed all had their distances, and hence their sizes, doubled. However, since these nearby galaxies had been used to calibrate the Hubble law for the determination of very distant galaxies, all the distances had in the end to be doubled, and hence so did the size of the Universe as calculated at that time.

A lot of effort has gone into correcting the position of the zero point. Apart from the work of W. Baade, there is also that of Shapley, Stebbins, Thackeray, Blaauw and Morgan, Gascoigne and Eggen, all before 1960, and of others more recently. We now have a much more exact relationship available which gives us the absolute magnitudes of typical Cepheids fairly closely. Figure 19 shows this relationship, or rather the two relationships, since the $C\delta$ and CW Cepheids form two groups lying along nearly parallel lines. In contrast, the RR Lyrae stars form an entirely separate group.

It is important to study stars with periods of about 1 day since we have seen that the ratios G_V and G_U made it clear that some of them are RR Lyrae stars and others are Cepheids of Population II. How can the boundary between these two groups be determined? For the moment it is difficult to say.

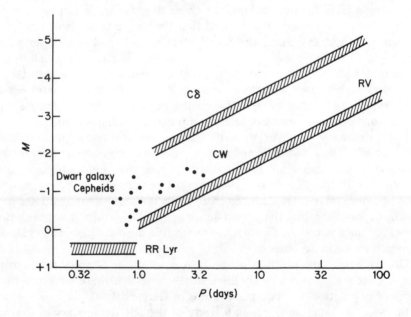

Figure 19.   The luminosity–period relationship

A large dispersion in the period–luminosity relationship can be seen. Is this a random scatter, due to inaccuracies in the absolute magnitudes, or is it real, the result of some stars departing significantly from the relationship?

The question is unresolved: undoubtedly, some of the dispersion is the result of imprecision in measurements, but an examination of Cepheids in the Magellanic Clouds and in certain nearby galaxies inclines me to the view that part of the dispersion is probably real.

DISTRIBUTION OF CEPHEIDS IN THE GALAXY

The 'classical' Cepheids (Cδ and Cδs) make up a typical Population I; most of them are located in the spiral arms of the Galaxy and form a very flat system, 90% of them being situated at less than 100 parsecs from the galactic plane and none at all at more than 300 parsecs. They are young stars, and some workers (Fernie, for example) have shown that many of them, mainly of long period, are associated with clouds of hydrogen in the galactic arms.

The situation is completely different for the CW stars. These are distributed throughout the galactic halo and their numbers decrease only very slowly as the distance from the galactic plane increases; they are still found at several thousand parsecs from this plane.

In 1960, M. Petit showed that the average distance to the galactic plane is 90 parsecs for classical Cepheids and 610 parsecs for W Virginis stars. The mean radial velocity is 15 km/s for Cδ and 36 km/s for CW.

An interesting phenomenon was revealed by Van den Bergh in 1958 and studied in more detail by Petit. Classical Cepheids are distributed differently according to their period; those with the shortest periods occur in large numbers in the outer arms of the Galaxy, in other words, those of the Cassiopeia, Orion, Canis Major-Puppis regions. Those with long periods often occur in the arms nearer the galactic centre (Carina-Centaurus, Vela, Sagittarius).

It is the same with the stars of Population II: the short period stars are to be observed towards the galactic longitude of 180°, or towards the anticentre, whereas the CW stars located in the direction of the galactic centre (Sgr-Sco) have long periods, often more than 5 days. This is illustrated in Figure 20. There is thus a clear relationship between the period and the galactic location.

Typical cepheids, or those with small amplitudes, are often associated with galactic clusters; thus, U Sgr is in the cluster Messier 25. There is one curious case that should be mentioned: in the open cluster NGC 7790, two Cepheids are known, CE and CF. However, CE Cas is a double star and the two components, separated by 2″, are both Cδ Cepheids with periods of 5.1411 and 4.4793 days.

Some long period Cepheids belong to O associations, and are very young objects. This is the case with U Car, a bright Cepheid of long period (38.8 days), which belongs to the stellar association Car OB 1.

About 35 CW stars are known in globular clusters, their light curves

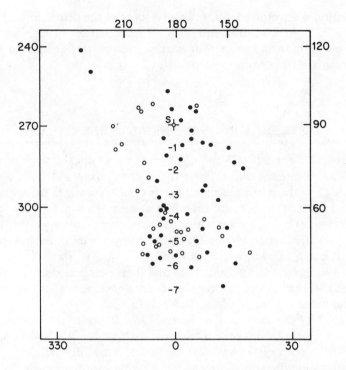

Figure 20. Distribution of population II Cepheids in the Galaxy. Stars with periods between 1 and 2.5 days are plotted as dots and those with periods of more than 12 days as open circles. S indicates the Sun and the distances are expressed in kiloparsecs

resembling those of the CWs in the galactic halo. The periods are of the same order, with two modal values: one around 1–2 days and the other between 10 and 20 days, only a few having periods between 3 and 10 days.

The mass of Cepheids is not known with great precision since no direct determination is available. It is estimated that it is about four solar masses for those with the shortest periods and about ten solar masses for those with the longest. Cepheids of Population II are distinctly less massive.

Like all supergiants, these stars must be evolving rapidly. A. Cox has calculated that the Cepheid state should not last longer than 10 million years.

CEPHEIDS IN OTHER GALAXIES

Cepheids have been observed in all the nearby galaxies, particularly in M31 (Andromeda), but the two formations that undoubtedly exhibit the richest collections are the satellite galaxies of our own, known as the Magellanic Clouds.

Systematic research on the Cepheids in these galaxies began at the Harvard Observatory at the beginning of the century and has continued up to the present day. It has been extremely fruitful in that more than 1100 Cepheids (in addition to numbers of other important variables) are now known in each of the Clouds.

It is interesting to compare the distribution of periods in the Galaxy with that in the two Clouds. This is illustrated in Figure 21, in which all the Cepheid types are included. For the Galaxy, as we have said, there are two modal values of the period, one between 4 and 6 days for classical Cepheids and the other between 10 and 15 days for the remainder, which contain a significant proportion of W Virginis stars.

The Small Magellanic Cloud (S.M.C.) also has two modes in the frequency distribution of the periods, one between 1.5 and 2 days and the other between 2.5 and 4 days. The periods are thus much shorter, on average, than those in our Galaxy, although there are a few very long ones of more than 100 days.

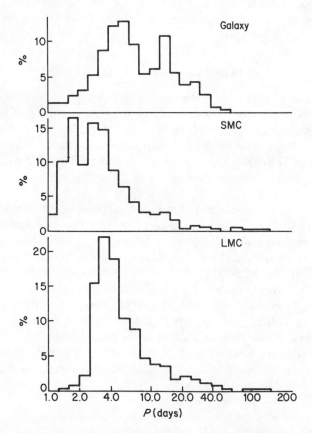

Figure 21. Distribution of periods (in %) for 701 Cepheids in the Galaxy, 1125 in the Small Magellanic Cloud (SMC), and 1110 in the Large Magellanic Cloud (LMC)

The distribution is very different in the Large Magellanic Cloud (L.M.C.), where there is only one modal value of between 3 and 4 days. There are also, however, a few very long periods.

These differences in distribution reflect differences in structure. The Large Cloud is mainly composed of Population I stars. There are many stars with short periods in the central region, whereas those with long periods tend to be distributed around the periphery. The Small Cloud, on the other hand, has both population types. The short period stars are observed to be towards the outside, as in our Galaxy, while the long periods are concentrated in the central regions.

In the Andromeda galaxy, Cepheids have been observed mainly in the arms, since it is difficult to resolve individual stars in the nucleus. The periods are often long, but it is true that at that distance from our Galaxy the short-period stars will be very faint and thus difficult to detect. It has been noted, however, that at an angular distance of 15' from the centre, the period occurring most frequently is 13.8 days; at 50' it is 7.0 days and at 96' it is 3.9 days. The period thus tends to decrease from the centre of the galaxy to its edge.

VARIABLES IN DWARF GALAXIES

The name 'dwarf galaxies' is given to small spherical formations that can be observed only at relatively short distances because of their low luminosity. The first two, discovered by H. Shapley in 1935, are called Fornax and Sculptor after the constellations in which they are observed. Their distances are 78 000 and 188 000 parsecs and their diameters 2400 and 6100 parsecs, respectively. Others are smaller still, like Draco which is less than 1000 parsecs across.

RR Lyrae stars have been found in dwarf galaxies, which is not surprising since they are similar to globular clusters. The numbers of variables in them is large: more than 600 in Sculptor, 260 in Draco and 200 in Leo II. The periods are of the same order as in the Galaxy, between 0.56 days (Sculptor) and 0.64 days (Ursa Minor).

Since 1975, some Cepheids with short periods ranging from less than 1 day to nearly 3 days have also been discovered. About 25 of these are known, 10 of them in Leo II. The interest in these Cepheids, which some writers call 'abnormal,' is that they are located in the period–luminosity relationship *between* Populations I and II and above the RR Lyrae stars, as is shown in Figure 19. They thus seem to form an additional type of Cepheid , but there is still very little known about them.

Chapter 3

Long-period Variables

There are a great many long-period variable stars, more than 5000 in the General Catalogue.

The typical star in this group is o Ceti or Mira, already mentioned as the first variable star discovered (Fabricius 1596). The first determination of its period was made by Bouillaud in 1667. He found a value for P of 333 days but also noted that this was not constant and that the brightness was not the same from one maximum to another. These results were confirmed by W. Herschel at the end of the eighteenth century, and later in 1859 by Argelander who, helped by the previous results, found an average period of 331 days 15 h. This is an accurate value for the mean period and is the one always adopted.

Mira reaches a magnitude of 2 at some of its maxima but at others it barely reaches 5. The irregularity also occurs at the minima, with the magnitude varying between 8.3 and 10.1. The maximum amplitude is 8.1 magnitudes and the average 6.5 magnitudes.

The period is subject to appreciable fluctuations: the mean value is 331.6 days with observed extremes of 310 and 370 days. We shall see that this behaviour is common to all long-period variables.

In the particular case of Mira, the problem is complicated by the fact that the star is double (or perhaps triple). The existence of the companion was suspected by Joy from spectral anomalies and was confirmed in 1923 by Aitken. This companion has a Beq spectrum which resembles that of P Cyg, and is itself a variable (V magnitude 9.6–12.0) with a designation VZ Cet. Its variability was detected in 1939 by P. Baize and later confirmed by many others.

The orbits of the pair have been the subject of many studies, but they are still not very well determined since the arc covered by the companion since its discovery (13°) is too small. Baize in 1980 found an orbital period of 400 years for it, with a mean angular separation of the components of 0.85" and an eccentricity of 0.66. In addition, he assumed that a third body exists with a period of revolution of about 29 years.

The maximum absolute visual magnitude of Mira is about -2 and that of its companion VZ Cet about $+4.5$, which means that the latter can be classified as a sub-dwarf.* The diameter of Mira, measured several times by interferometric methods, is $0.053''$, which corresponds to 4 astronomical units. If it took the place of the Sun, Mira would extend well beyond the earth's orbit and even include that of Mars.

SPECTRA, PERIODS AND AMPLITUDES

Long-period variables have spectra of type M, C or S. Type M are by far the most numerous, the others making up less than 10% of the total. Figure 22 shows the light curves for Mira (M spectrum) extending over 1000 days, and for χ Cyg (S spectrum) and V Oph (C spectrum).

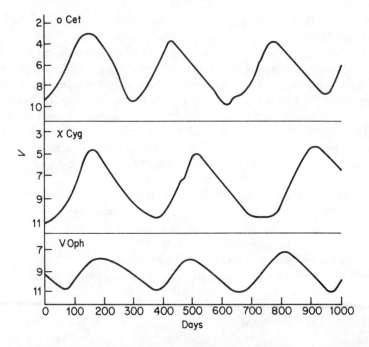

Figure 22. Light curves plotted over 1000 days of o Cet (Mira), an Me spectrum variable with a period of 331.6 days, of χ Cyg (Se spectrum, 406.8 days) and of V Oph (Ce spectrum, 298.0 days)

Many long-period variables have an M spectrum showing intense emission lines (denoted by Me), particularly of hydrogen. These belong to the luminosity classes II and III, and thus originate from giants.

---

* The dynamic parallax calculated by P. Baize leads to different results: absolute magnitudes -3.1 and $+3.5$, masses 15.7 and 4.0 solar masses. The period calculated provisionally (400 years) may not be accurate.

Table 11 shows the details of twenty long-period variables with Me spectra, chosen from the brightest and best known of them, and Tables 12 and 13 give details of nine stars with Se spectra and six with C spectra, respectively (as before, V_{max} is the maximum visual magnitude, ΔV is the amplitude and D the asymmetry).

Table 11.   Long-period variables with Me spectra

Designation	V_{max}	ΔV	P (days)	D	Spectra	
S Car	4.5	5.4	149.5	0.51	K7e–M4e	III
T Her	6.8	6.8	165.0	0.47	M2e–M5.5e	III
R Boo	6.7	6.1	229.5	0.46	M3e–M6e	III
R Dra	6.9	6.1	245.6	0.45	M5e–M7e	III
R Tri	5.5	7.1	266.4	0.44	M4e–M9e	III
R Aql	5.7	6.3	293.0	0.42	M5e–M8e	III
R UMa	6.7	6.7	301.8	0.39	M3e–M6e	III
R Leo	4.4	6.9	312.6	0.43	M6.5e–M9e	III
o Cet	2.0	8.2	331.6	0.36	M5e–M7e	III
X Oph	5.9	3.3	334.2	0.53	M5e–M7e	III
R Ser	5.7	8.7	356.8	0.41	M6e–M8e	III
R Cnc	6.2	5.6	361.7	0.47	M6e–M8e	III
U Ori	5.3	7.3	372.4	0.38	M6e–M8e	III
R LMi	6.3	6.9	372.3	0.41	M7e–M8e	III
S Vir	6.3	6.9	377.9	0.45	M6e–M7e	III
T Cep	5.4	5.6	387.8	0.54	M5e–M9e	III
R Hya	3.0	8.0	388.0	0.48	M6e–M8e	III
R Hor	4.7	9.6	402.7	0.40	M5e–M7e	II–III
R Cas	5.5	7.5	431.0	0.40	M6e–M8e	III
R Aur	6.7	7.0	458.4	0.51	M7e–M9e	III

Table 12.   Long-period variables with Se spectra

Designation	V_{max}	ΔV	P (days)	D	Spectra
R Cam	7.9	6.5	270.1	0.45	S2e–S8e
W Cet	7.1	7.5	351.1	0.50	S6e
R Gem	6.0	8.0	369.6	0.36	S3e–S7e
R Lyn	7.2	6.8	378.9	0.44	S3e–S7e
W And	6.7	7.8	396.7	0.42	S6e
χ Cyg	3.2	10.9	406.8	0.41	S7e–S10e
R And	6.0	8.9	409.0	0.38	S6e
R Cyg	6.5	7.7	426.5	0.35	S4e–S7e
S Cas	7.0	7.3	611.0	0.40	S3e–S6e

Table 13.   Long-period variables with C spectra

Designation	V_{max}	ΔV	P (days)	D	Spectra
R CMi	7.4	4.2	337.9	0.48	Ce
R Ori	9.1	4.3	378.0	0.40	C8e
WX Cyg	8.8	4.4	410.6	0.48	C9e
R Lep	5.5	5.0	432.5	0.55	C7e
U Cyg	6.7	4.7	464.6	0.48	C7e–C9e
S Cep	7.4	5.5	487.4	0.55	C7e

The distribution of periods is given in Figure 23 for 4547 stars of all the types together, while separate histograms are given in Figure 24 for those with known spectral types, that is, for 1356 M-type stars, 69 S-type stars and 79 C-type stars. It can be seen that M stars have periods from 80 to more than 700 days, with a modal value of 280–290 days. The S stars have longer average periods, with a mode at 360 days, and C stars have their maximum frequency at 450 days.

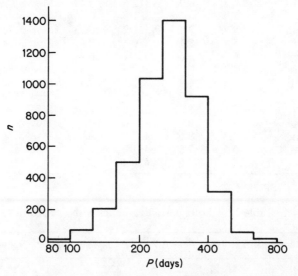

Figure 23.   Overall distribution of the periods for 4547 variables of the Mira type

It is interesting to examine the amplitudes of variation. Figure 25 shows the distribution of these for 445 variables whose visual amplitude is known. Of these, 10 are M-type stars without emission, 357 are of Me spectral type, 48 are S-type and 35 C-type. The M stars without emission have an amplitude of 4–5 magnitudes, the Me stars a larger amplitude of 6.3 magnitudes on average, and the figure reaches 7.2 for the S stars and less than 5 for C stars.

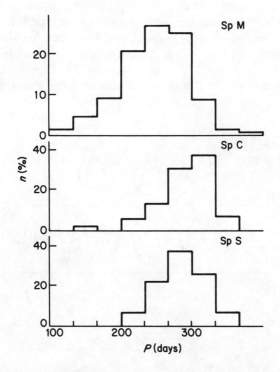

Figure 24.   Distribution of periods for 1356 long-period variables with M spectra, 79 with C spectra and 69 with S spectra

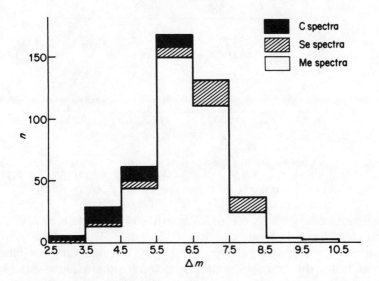

Figure 25.   Distribution of amplitudes Δm of 445 long-period variables

58

Many studies have been made of the light curves, particularly by Ludden-dorf, Campbell and Gaposchkin. As regards the M stars, the ratio D defining the asymmetry lies between 0.4 and 0.5, and Campbell has shown that vari-ables with periods less than 200 days have much more rounded light curves than those with longer periods. This is illustrated in Figure 26. For C stars, D is about 0.5; that is, the rise to the maximum takes the same time as the decline.

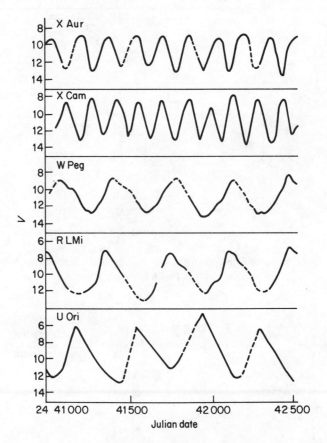

Figure 26.   Light curves for five long-period variables with Me spectra: X Aur (P = 164.0 days), X Cam (143.5 days), W Peg (344.9 days), R LMi (372.3 days) and U Ori (372.4 days). From J. Aubaud (AFOEV observations)

A very interesting case is that of R Aquarii, which normally varies between magnitudes 5.8 and 11.5 with a period of 386.3 days. This is a binary, formed from a red giant of spectral type M7 III and a small very hot blue star responsible for the emission, which is itself surrounded by a vast gaseous envelope. An analysis of the light curve by J. Mattei and J. Allen has shown that it is produced by eclipses due to the passage of the gas cloud in front of the red giant. During this time, the amplitude decreases rapidly, with the star varying between magnitudes 9 and 11.5. This occurred in 1890, 1934 and then in 1978, giving an orbital period of 44.5 years. The eclipses are, moreover, very long-lasting: the perturbations produced in the variations extend over 8 cycles, that is, 8.5 years. The last 'minimum' started in 1974 and ended in 1982.

Binaries of this type are probably frequent among the long-period variables. They are similar to symbiotic stars (of the Z Andromeda type) that we shall deal with later.

To end this description of the curves, we quote a very odd variable star, R Cen (Figure 27). This exhibits two kinds of minima, one shallow and the other deep, which alternate regularly. The period defined by the deep minima is 548 days. Several workers have emphasized the similarity of R Cen to the RV Tauri stars, which are dealt with in the next chapter.

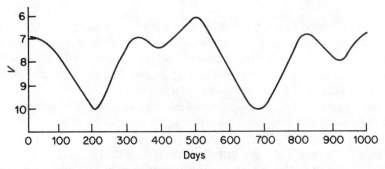

Figure 27.   Light curve for R Cen, a peculiar long-period variable with a double maximum ($P = 543.0$ days)

VARIATIONS IN PERIOD

We have seen that all these stars have variations in their periods, but several different cases must be distinguished.

The most general type of variation is an oscillation around a mean value of about 10% which takes several years or decades in all. Over long time intervals, the mean value itself shows no change or almost no change. A constant mean period can then be quoted, and this is the case for Mira, χ Cyg, R Tri and many other stars.

For some stars, the oscillation about the mean period can take a very long time. This can be seen in the $O-C$ diagram relating to SY Herculis (Figure 28), where it extends over more than 200 maxima.

Several stars show another effect: a continuous decrease in the period. This is particularly so with R Hya, a star known since 1662. In 1704, J. P. Maraldi estimated its period as 547 days, but Pigott in 1784 indicated that it was no more than 487 days. In 1880, it had fallen to 432 days, in 1920 to 404 days and in the last 20 years to 388 days. The shrinkage in the period,

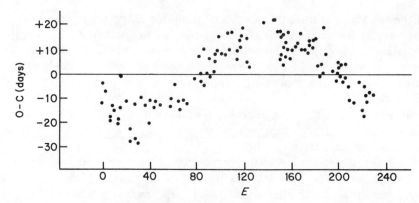

Figure 28. O–C diagram for SY Her (*P* = 116.8 days). *E* is the number of maxima measured from a point taken as origin. (*From P. Ahnert*, Mitt. Veränderlichen Sterne, 7, No. 5, 1976. (*Reproduced by permission of Sternwarte Sonneberg, GDR*)

at first fairly rapid (60 days in 80 years) has now slowed: over the last 270 years, the reduction has been on average only 0.6 days per year. The same effect was observed in R Aql: discovered by Argelander in 1856, it then had a period of 350 days whereas now it is 293 days. The decrease has been less rapid than that of R Hya, but is still considerable.

Known cases of this kind are still few, but the fact is that because the cycles are long and irregular, a considerable time of as much as a century or more is needed to tell the difference between a cyclic variation such as that of SY Her and a secular one such as that of R Hya.

Most of the variables we observe were not known 100 years ago, so that another question can be raised which does not yet have an answer: could the variations like those of R Hya and R Aql not be cyclic but with a cycle so long that we cannot yet detect it?

PHYSICAL CHARACTERISTICS

The spectra of the M and S stars show that they are low-temperature stars. With *o* Ceti, for example, it has been observed that the colour temperature fluctuates between 2640 and 1920 K and for R Tri between 2400 and 1950 K.* Some are even cooler.

The appearance of the spectra changes completely in the course of a cycle: the emission lines, particularly those of the hydrogen Balmer series, of silicon and of iron, which are intense near maximum brightness, decrease in intensity as the brightness decreases and disappear almost entirely at the minimum, only to reappear in the rising phase of the next cycle. The metallic absorption lines become the most important ones at the middle of the

---

* Stellar temperatures are expressed on the absolute (Kelvin) scale, denoted by K. Absolute zero (0 K) is −273 °C.

decline. As regards the absorption bands (mainly titanium oxide for the M stars, zirconium oxide for the S stars and to a lesser extent lanthanum oxide), they are weak at the maxima but become very strong at the minima. We shall not describe these spectra in more detail, but merely indicate an interesting point: traces of several isotopes of technetium have been found in some S stars. Now, all the isotopes of technetium are unstable, and their very short lifetime shows that they can only be formed in the atmosphere of a star.

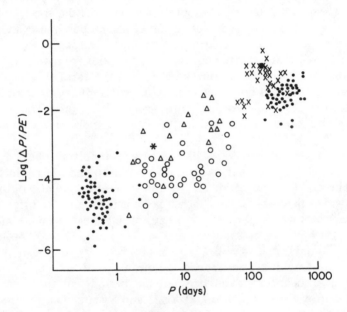

Figure 29. Variations in the periods of pulsating stars. The dots on the left refer to RR Lyr stars; the circles to Cδ stars; the triangles to CW stars; the crosses to semi-regulars and the dots on the right to Mira stars. (*From D. Hoffleit*, Inf. Bulletin of Variable Stars, *No. 1131, 1976. Reproduced by permission of the Konkoly Observatory Budapest*)

It should be noted that at maximum brightness, the emission lines have a different velocity from that of the absorption lines, by about 10–20 km/s. At the minima, on the other hand, the radial velocities are almost identical.

These stars are very red and their spectra very intense in the infrared region. Photoelectric measurements carried out by various workers such as Lockwood and Wing or Barnes have shown that, at the maximum, the colour index $V-I$ is very high: +5.0 for o Ceti, +8.5 for T Cep and more than 10 in some cases. It follows that in the infrared, Mira reaches a magnitude of -2 and χ Cyg a magnitude of -5.

Because the changes in the colour index are so great from the maximum to the minimum, the amplitude is much smaller in the infrared than in the visible region: it is only 1.1 magnitude for *o* Cet (instead of 8.2 in *V*), 0.55 magnitude for T Cep (instead of 5.6) and 2.0 magnitudes for χ Cyg (instead of 10.9).

The C stars are completely different from the M and S stars. They are often even redder than the M stars: for V Aql, the $B-V$ index is +4.0 and the $U-B$ index is about +6.5. This means that if the star has a magnitude of 7 in *V*, it has one of 11 in *B* and 17.5 in *U*! These stars are, however, hotter than the M or S stars. The distribution in the continuous spectrum corresponds to that of a G star for the C0 class, and to a K or M star for the latest C class stars. The reason for this is as follows: the spectra of C stars show many absorption bands due to molecular carbon and cyanogen, and these are responsible for the strong reddening of the star.

One anomaly that is worth pointing out as a curiosity involves the isotopes of carbon. As well as the normal isotope ^{12}C, the earth's atmosphere also contains ^{13}C and so a bond between a ^{12}C and a ^{13}C atom can be formed in addition to that between the normal pair of ^{12}C atoms. The ^{13}C isotope is very scarce on the earth and in the Sun, but it is very abundant in C stars.

The long-period variables, like the semi-regulars, obey a period–luminosity relationship that was studied by Wilson and Merrill in 1942 and by Osvalds and Risley in 1961. The relationship is the converse of that in the Cepheids: in other words, the shortest periods belong to the brightest stars. Thus, variables with a period of 175 days have a mean absolute magnitude of −2.7; for a period of 250 days, $M_V = -1.7$; for 350 days, $M_V = -0.5$; and for 500 days, $M_V = +0.9$. These, however, are for mean values and there is considerable scatter around the means.

There is also a correlation between the period and the spectral type, pointed out by Gaposchkin (1938) for C and S stars. The mean spectrum passes from M3 for periods of about 150 days to M5 for periods of 270 days and to M7 for 400 days, all with considerable scatter about these mean values.

THE MECHANISM OF THE VARIATION

Why do these long-period variables have such a large amplitude of variation in brightness? Their radius changes by about 10% over a cycle, yet we have seen that in Cepheids with changes of the same order the amplitude of brightness is only about 1 magnitude. In long-period variables, there is a much greater amplitude: up to 11 magnitudes for χ Cyg, which means that it is 20 000 times brighter at its maximum than at its minimum.

Several theories have been advanced to explain this anomaly. Thirty years ago, P. W. Merrill pointed out that the atmospheres of these stars were rich in titanium oxide molecules. When the star is hot, the molecules dissociate and when it cools, the atoms recombine to form a highly absorbant 'cloud' in the stellar atmosphere.

More recently, it has also been pointed out that titanium oxide is probably not solely responsible for the effect. Observations made with infrared satellites have shown that other molecules are present in the atmospheres, such as oxides of carbon, cyanogen and water, and that there is also 'dust' formed from metallic silicates. All these molecules must dissociate when the star becomes hotter and recombine when it cools, to make the atmosphere very opaque. Calculations made in recent years have been able to explain the considerable amplitudes of the variations in brightness.

Recent work in the infrared and radio regions of the spectrum has contributed to a better understanding of the effect. The pulsation arises in the deepest layers of the star and is propagated to the outer regions which are less and less dense as the surface is approached. In these less dense parts of the star, the pulsation gives rise to a shock wave which heats up and thus dissociates the molecules and dust particles. After the shock wave has passed, the temperature falls and the situation reverts to what it was before: the molecules recombine and the opacity becomes very high.

It is known that some of these stars are binaries, consisting of a red giant and a very hot dwarf. As we have seen, this is the case with Mira and several others. In such a pair, the hot component is not very bright and is difficult to detect. It can, however, play an important role in determining what is observed; these companions are generally surrounded by a gaseous cloud at a high temperature and there must be an interaction with the very tenuous atmosphere of the red giant. The complete theory of this effect has still to be worked out in detail, but research work is in progress at the moment, in both the theoretical and observational fields.

GALACTIC DISTRIBUTION

Long-period variables do not form a homogeneous group. Stars with periods of less than 200 days belong to Population II: they are found in the halo, in the direction of the galactic centre and in globular clusters. They have considerable radial velocities, from 65 to more than 200 km/s. Those with periods greater than 200 days form a type of Population I described as intermediate; they are to be found up to 1000–1200 parsecs from the galactic plane and their radial velocities range from 15 to 50 km/s, depending on the period.

As regards the S spectra stars, they are very young, as we have said, and they form a flat system typical of Population I with distances from the galactic plane of less than 100 parsecs.

The galactic distribution of long-period variables has been particularly studied by Romano and Di Tullio. Figure 30 shows the distribution of the periods as a function of galactic position. In the direction of the galactic centre which, as expected, includes numerous Population II stars, the proportion with periods below 200 days is high, and those with periods above

300 days are relatively few. It is the converse in the direction away from the centre, where the modal value occurs around 350 days with a still sizeable proportion having periods greater than 400 days.

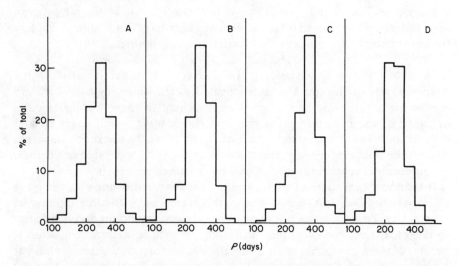

Figure 30. Distribution (in %) of the periods of Mira stars with respect to galactic position. A: For all 4450 long-period variables. B: For 725 variables of high galactic latitude. C: For 115 stars in the direction away from the galactic centre. D: For 965 stars in the direction of the galactic centre

We end by emphasizing that, like the RR Lyrae stars but unlike the Cepheids, long-period variables are not good indicators of distance, since their period–luminosity relationship is too ill-defined. However, their great luminosity makes them visible at very large distances (they are known in several nearby galaxies, particularly in the Magellanic Clouds). Moreover, the presence of intense emission lines enables their radial velocities to be easily determined. They are thus used in kinematic and dynamic studies of galactic structure.

Chapter 4

Semi-regular and Irregular Variables

The appropriate name 'semi-regular variables' is given to those red pulsating stars which are distinguished from the long-period variables by an often more complex light curve and by a smaller amplitude rarely exceeding 2 magnitudes. The name 'irregular variables' is given to those stars whose luminous fluctuations are too independent of time for it to be possible to define a cycle. It should be said at once that the separation between semi-regulars and irregulars is slightly arbitrary, since it is difficult to say where the semi-regularity stops and the irregularity begins. Some stars behave at times like semi-regulars and then become irregular, and vice versa.

The General Catalogue divides the semi-regulars into four sub-types: SRa, SRb, SRc and SRd. Sub-types SRa and SRb include giants with M, S or C spectra; sub-type SRc includes M supergiants; and SRd supergiants with G or K spectra. In the same way, irregulars include two sub-types: Lb are giants with M, C or S spectra and Lc are the M supergiants.

We describe all these sub-types in this chapter and also include in it a group normally considered separately, but too secondary to have a chapter devoted entirely to them. These are the RV Tauri stars, whose variations follow their own peculiar path. This is not at all the case for the semi-regulars and irregulars, which are red giants very similar to the long-period variables. Yet, if the mechanism behind the variations is on the whole very much the same in these various cases, it is difficult to understand why the fluctuations in luminosity of the semi-regulars and regulars are weaker and more irregular than those of the long-period variables.

RV TAURI STARS

RV Tauri stars are sometimes called 'semi-regular Cepheids,' where the name Cepheid is used in the long-used general sense of yellow–orange pulsating stars with medium periods. They have a relatively regular cycle, but the magnitudes of both the maxima and minima are not constant from one cycle to the next.

The light curve is curious: one maximum is followed by a shallow minimum and then by a second maximum, which is itself followed by a deeper minimum. It is convenient to call a 'period' the time between two deep minima. Superposed on these fluctuations is a 'principal cycle' of period P_b separating the fluctuations of maximum amplitude. P_b is in fact a modulation period. For the typical star RV Tauri, the normal period is 78.7 days and that of the 'principal cycle' is 1224 days.

Two sub-classes have been created, RVa and RVb, depending on the regularity of the magnitude at the maximum. However, there do not seem to be great differences between the physical states of the two groups.

RV Tauri stars are not very common and there are only about 100 of them in the Galaxy. Table 14 gives the properties of eight of the better observed ones, while Figure 31 shows the light curves of two of them: U Mon (RVa) and R Sct (RVb). The periods run from 30 to more than 200 days, with a fairly distinct modal value between 60 and 80 days. Periods greater than 150 days are rare.

Table 14. RV Tauri stars: the column headed P_b gives the mean period of the principal cycle

Designation	B_{max}	ΔB	P (days)	P_b (days)	Spectra	Sub-type
SU Gem	10.7	2.4	50.12	690	K0–M3	RVb
TU Oph	10.43	1.85	61.08		G2e–K0	RVa
R Sge	9.46	2.00	70.59	1112	G0–G8 Ib	RVb
AC Her	7.43	2.51	75.46	9400	F2p–K4e Ib	RVa
RV Tau	9.90	3.45	78.70	1224	G2e–K3p Ia	RVb
TW Cam	10.4	1.2	86.26		F8–G8 Ib	RVa
U Mon	6.1	2.0	92.26	2320	F8 IIe–K0 Ib	RVb
R Sct	5.87	1.99	140.2		G0e–Sa–K0p Ib	RVa

These stars are supergiants, with spectral types from F to K and sometimes M, and with intense emission lines when near the maximum. They are very bright, with absolute magnitudes from -3 to -4. They belong to Population II and are encountered in globular clusters (three are known in Messier 5) and in certain galaxies (three or four are known in Messier 31).

On the whole, these stars are similar to W Virginis stars, in spite of having light curves that are significantly different. It can be seen from Figure 19 that they follow the period–luminosity relationship of the Population II Cepheids.

These stars are of great size but small mass and hence of very low density. Tsesevitch put forward the theory in 1955 that the pulsation, which starts in the hotter layers, creates a shock wave that arrives at the very tenuous outer layers of the star and amplifies the variations in the radius. An exact model

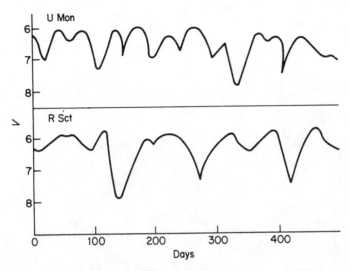

Figure 31.   Light curves for two RV Tauri stars: U Mon (type RVa), $P = 92.3$ days; R Sct (type RVb), $P = 140.2$ days

explaining all the observations has still to be constructed, but it is certainly known that the shock wave, as in the long-period variables, has the effect of encouraging the formation of emission lines near the maximum.

THE SUB-TYPE SRa

These stars have light curves (Figure 32) that make them resemble long-period variables with a smaller amplitude and greater symmetry; the factor D is approximately 0.5. The period is certainly not very regular, but its irregularity is hardly greater than that of the long-period variables. In practice, therefore, it is difficult to fix the boundary separating these two groups. In a somewhat arbitrary way, it has been decided to classify those stars with amplitudes greater than 2.5 magnitudes as long-period variables and those with amplitudes smaller than 2.5 as SRa semi-regulars.

A large number of SRa stars are known: the General Catalogue and its supplements lists at least 1200. The spectra may be type M (with or without emission), type C or rarely type S. Out of 470 stars of known spectra listed in the Catalogue, there are 385 with M spectra, 68 with C and 17 with S. Table 15 gives the properties of 15 of the brightest of these. The magnitudes and ranges were obtained photographically (pg) or photovisually (pv).

The periods range from 40 to nearly 1000 days. However, a distinction should be made between the spectral types. For type M stars (Figure 33), there is one modal value between 150 and 200 days with a small secondary peak between 250 and 300 days. In contrast, the C and S stars have only one well marked peak at 350 days.

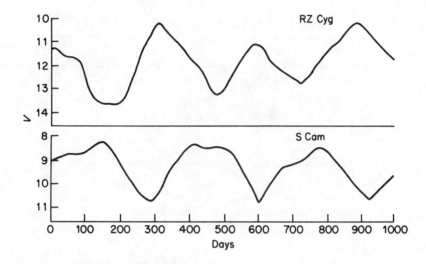

Figure 32.   Light curves for two SRa stars: RZ Cyg (spectrum M5 IIIe), $P = 275.7$ days; and S Cam (spectrum C7e), $P = 326.4$ days

Table 15.   Variables of sub-type SRa

Designation	m_{max}	Δm	P (days)	D	Spectra
V 450 Aql	7.0	1.9pg	64.2	0.52	M5 IIIe
T Cen	5.5	3.5pv	90.6	0.47	K0–M4 IIe
Z Aqr	9.5	2.5pg	136.0	0.33	M1e–M3e
RU Vul	8.8	3.4pv	156.3	0.47	M3.5 IIIe
V CVn	6.8	2.0pv	191.9	0.50	M5 IIIe
RU Cyg	9.2	2.4pg	234.4	0.50±	M6 IIIe
V Boo	7.0	4.3pv	258.2	0.49	M5.5 IIIe
RY UMa	8.1	1.6pg	311.2		M2e–M3e III
R UMi	8.8	2.2pv	324.4	0.50	M7 IIIe
W Hya	7.7	3.9pg	382.2	0.50	M8–M9e
WZ Cas	9.4	2.0pg	186.0	0.58	C9.2
ST And	8.2	3.6pv	328.0	0.54	C4.3e
VX And	7.8	1.5pv	369		C4.2
RS Cyg	6.5	2.8pv	418.0	0.45±	C8e
RY Mon	7.7	1.5pv	466	0.43	C5.5

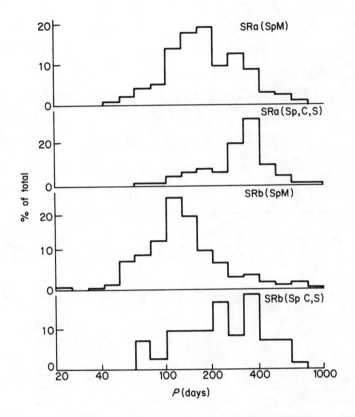

Figure 33. Distribution of periods in 385 SRa stars with M spectra; in 68 SRa stars with C spectra and 17 with S spectra grouped together; in 358 SRb stars with M spectra; and in 61 SRb stars with C spectra and 10 with S spectra grouped together

THE SUB-TYPE SRb

The characteristic feature of these stars is the existence of several superposed cycles. For example, W Cyg has at least two periods of 130.8 and 119.8 days. This produces fluctuations that are often complex, with irregular variations that may even show a quasi-constant brightness at times. It follows that there is a modulation period 8–15 times greater than the normal period: this is 900 days for g Her, 960 days for AF Cyg. 2450 days for W Ori and 2270 days for V Aql. This period corresponds, as always, to the maximum amplitude.

Figure 34 shows the light curves for three of these stars, AF Cyg, L₂ Pup and Z UMa, and Table 16 gives the properties of 15 of the brightest and best known SRb stars.

The General Catalogue lists about 700 variables of this sub-type. We shall see later that they may in reality be more numerous than SRa stars. Their spectra belong to types M, C and S; of 429 stars with known spectra, 358 are

70

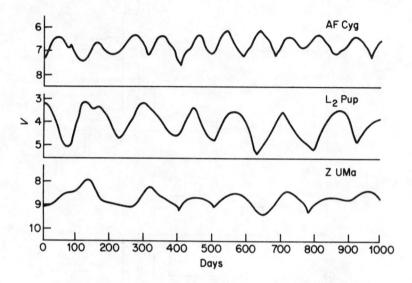

Figure 34.   Light curves for three SRb variables: AF Cyg (M5 IIIC spectrum, P = 94.1 days), L₂ Pup (M6e, 140.8 days) and Z UMa (M5 IIIe, 196.0 days)

Table 16.   Variables of sub-type SRb

Designation	m_{max}	Δm	P (days)	Spectra
R Lyr	3.9	1.1pv	46.0	M5 III
EU Del	6.0	0.9pv	59.5	M5 II
g Her	5.7	1.5pg	70	M6 III
AF Cyg	7.4	2.0pg	94.1	M5 IIIe
W Cyg	6.8	2.1pg	130.8	M4e–M6e
X Mon	6.9	3.1pv	155.7	M3–M4 IIIe
T Cet	7.6	1.1pg	158.9	M2 IIe
RT Mon	8.2	2.1pg	195.3	M3 III
V Aqr	7.6	1.8pv	244.0	M6e
R Dor	5.9	1.0pv	338	M7 IIIe
Y CVn	8.2	1.8pg	158.0	C5.4
W Ori	8.6	1.5pg	212	C5.4
UU Aur	8.2	1.8pg	295	C4
V Aql	10.6	1.9pg	353	C5.4
R Scl	9.1	3.7pg	363.1	C6.4

type M, 61 are type C and 10 are type S. The periods have a similar range to that of the previous sub-type, from 30 days to nearly 1000 days, with a well marked peak frequency of occurrence of 100–125 days for M stars, and with two such peaks, one around 200–250 days and the other 350–400 days for C and S stars. However, this should not conceal the fact that there are many M stars with short periods of 30–40 days.

THE SUB-TYPE SRc

Sub-type SRc stars are all M supergiants and extremely bright with absolute magnitudes of −5 to −6. The typical star is μ Cep, discovered by W. Herschel in 1782, which has an amplitude of 1.5 magnitudes and very complex variations (Figure 35); Hasseinstein has found three periods of 730, 904 and 4900 days, while Balasoglo found four, of 700, 900, 1100 and 4500 days.

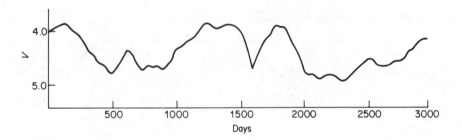

Figure 35.   Light curve over 3000 days for μ Cephei (type SRc)

However, by applying statistical methods, Ashbrook, Duncombe and Van Woerkom reach a very different conclusion; according to them, the Balasoglo cycles are only apparent and the variations are completely irregular. Klyus returned to the problem in 1978 and found a principal cycle of 728 days corresponding to the rotation period of the star, and secondary cycles of 364, 182, 122, 90 and 73 days, which are respectively 1/2, 1/4, 1/6, 1/8 and 1/10 of the principal cycle.

At the moment, it is difficult to come to a decision about this. Only one thing is certain: the variations are complex and, if they are periodic, there are superposed cycles. Thus, for SS And, periods of 152, 650 and 3800 days have been found; for U Lac, 150, 550 and 2900 days; for T Per, 326 and 3800 days; for S Per, 810 and 916 days; and so on.

There are very few of these stars, which would be expected since they are very bright. About 40 are known, and details of five of these are given in Table 17. It should be pointed out that some are particularly bright, such as α Ori (Betelgeuse) and α Sco (Antares).

Table 17. Variables of sub-type SRc

Designation	V_{max}	ΔV	P (days)	Spectra
α Her	3.06	1.0	130	M5 Ib II
TV Gem	6.6	0.8	182	M2 Iab
μ Cep	3.65	1.5	730	M2 Iae
α Sco	0.88	0.9	1733	M1 Iab
α Ori	0.42	0.9	2070	M2 Iab

These stars form a typical Population I and are present in some of the open clusters. Thus, about ten (more than a quarter of those that are known) are to be found in the double cluster of Perseus (h and χ Per). Table 18 gives the properties of this group together with those of three other supergiants which are considered to be irregulars of type Lc.

Table 18. SRc and Lc supergiants in the double cluster h and χ Per

Designation	V_{max}	$(B-V)_{max}$	ΔV	Spectra	Type	P (days)
S Per	7.9	2.71	3.6	M3e-M6e Ia	SRc	825
T Per	8.2	2.33	0.9	M2 Iab	SRc	260
W Per	8.7	2.3	3.1	M3-M5 Ia	SRc	465
RS Per	7.9	2.24	1.2	M3-M4.5 Iab	SRc	—
SU Per	6.8	2.19	1.1	M3.5 Iab	SRc	533
XX Per	7.7	2.14	0.8	M3.5 Ib	SRc	214
YZ Per	7.7	2.33	1.3	M2.5 Iab	SRc	378
AD Per	7.4	2.28	1.0	M2.5 Iab	SRc	330
BU Per	7.9	2.47	1.7	M4 Ib	SRc	365
FZ Per	7.5	2.27	0.9	M1 Iab	SRc	184
KK Per	6.6	2.22	1.2	M1-M2 Iab	Lc	—
PP Per	9.2	2.3	1.1	M0.5 Iab	Lc	—
PR Per	7.6	2.33	1.0	K5e M2 Iab	Lc	—

We have already mentioned that the boundary between the semi-regulars and the irregulars is not always very distinct, and this is certainly the case with the two groups SRc and Lc. All the variables in the double cluster are fairly bright (magnitudes 6–8) and with an appreciable amplitude of 1 magnitude or more and thus easy to observe. Nevertheless, they have until now been neglected by the associations of observers of variables stars.

THE SUB-TYPE SRd

This group was long confused with the RV Tauri stars. It is true that in both cases they are yellow supergiants with F5 to K spectra, but the radial velocities of the SRd type are much greater than those of the RV Tauri, often more than 100 km/s. The SRd stars are also less bright, with about ten of them known in globular clusters with absolute magnitudes of about −1.6. Finally, the light curves are different, as can be seen from Figure 36.

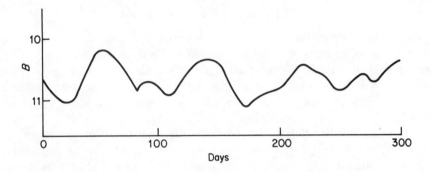

Figure 36.   Light curve of S Vul (type SRd)

There are very few of these stars, about 50 being known at present. Table 19 gives the properties of eight of them. The amplitude of the variations is fairly large, on average around 2 magnitudes. The period is not regular and nor is the light curve. In the case of UU Her, for example, the period was 71 days between Julian dates 2428080 and 2429100 with fairly flat maxima. Between 2429150 and 2430200 it increased sharply to 90 days, with the light curve then resembling that of an RV Tauri star. At some point the period then fell to 49 days, and ever since has been continuously changing, both increasing and decreasing.

Table 19. Sub-type SRd stars (pg in the Δm column refers to amplitudes estimated photographically, pv to those estimated photovisually)

Designation	m_{max}	Δm	P (days)	Spectra
TY Vir	8.00	0.32pv	50	G3p Ib
S Vul	9.8	1.3 pg	68.8	G0–K2 Ib
TZ Cep	9.0	2.2 pv	83.0	G6–K2e
UU Her	8.5	2.1 pg	71.90	F2 Ib–G0
AG Aur	10.0	1.6 pv	96.0	G0eIb–K0ep
SX Her	8.6	2.1 pv	102.9	G3ep–K0
WW Tau	9.0	3.0 pv	116.4	G2e–K2
TX Oph	9.8	2.3 pv	138	F5e–G6e

It is thus not easy to define the periods of these stars, even average ones. If those from the Catalogue are taken, it can be seen that they range between 30 and more than 200 days, with a modal value between 80 and 100 days. Some observers have occasionally announced periods of over 300 days but, in my view, these are still to be verified.

THE IRREGULAR VARIABLES

We have already mentioned that irregular variables are sub-divided into two groups: Lb, giant stars of spectral types M (sometimes K), S or C; and Lc, supergiants of type M. These stars are fairly numerous, with about 1500 being known. As in previous cases, the spectral type M is predominant: the General Catalogue lists 637 M stars, 18 S stars and 125 C stars. There are very few Lc stars (no more than about 40).

It is not possible to describe these stars since their variations in brightness have only one feature in common: an absence of periodicity. In some cases, the variations are slow and continuous, in others they are abrupt. A typical example is that of SW Cet, studied by Prager in 1941. From 1898 to 1920, this star showed only small variations of 0.5 magnitude. From 1920 until 1928, the brightness increased slowly by more than 1 magnitude. In little more than a year, it fell once more, and from 1930 to 1940 it showed only small oscillations with an amplitude of no more than 0.3 magnitude.

The subtype Lc includes some bright stars, such as ψ_1 and η Aur, o_1 CMa, β Gru and λ Vel. Their variations are often of small amplitude and are generally slow.

SEMI-REGULAR AND IRREGULAR VARIABLES
OF SMALL AMPLITUDE

Progress in photoelectric photometry has enabled several workers, particularly Eggen (1969), to reveal variations of small amplitude (from 0.05 to 0.30 magnitude) in a number of nearby bright red giants. With a limitation to stars of brightness greater than a visual magnitude of 6.5, about 150 new variables in this category have been discovered in the last few years. Table 20 gives details of ten of these stars as some indication of their nature.

Among these new variables are a large number of Lb type stars and semiregulars belonging to the sub-type SRb, which is what leads to the suspicion that SRb stars are perhaps more numerous than SRa.

In 1950, Hynek suggested that red variables of small amplitude could be more numerous than the long-period variables. Observations made during the last few years confirm this hypothesis. From the list of stars brighter than magnitude $M_V = 6.5$, I have taken a list of the confirmed red variables: a total of 309 stars. Amongst these are 27 long-period variables. Figure 37 shows that 75% of the bright red variables have an amplitude of less than 0.5 magnitude.

Table 20. Some bright semi-regular and irregular variable
stars with small amplitudes of variation

Designation	V	ΔV	Spectra	Type	P (days)
χ Aqr	5.02	0.23	M5 III	Lb	
η Aur	4.24	0.10	M3 III	Lc	
μ Gem	2.76	0.26	M3e III	Lb	
II Hya	4.85	0.27	M4 III	SRb	61
RX LMi	5.98	0.18	M4 III	SRb	150
δ_2 Lyr	4.22	0.11	M4e II	Lc	
ε Oct	4.96	0.40	M6 III	SRb	55
α Tau	0.75	0.20	K5 III	Lb	
λ Vel	2.14	0.08	K4 Ib II	Lc	
ψ Vir	4.7	0.11	M3 III	Lb	

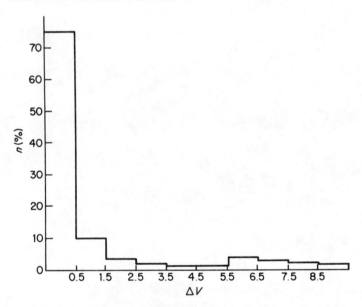

Figure 37. Distribution of amplitudes of 306 red variables brighter than
magnitude 6.5

GALACTIC DISTRIBUTION

We shall only give a brief indication here of the distribution of these various
categories of star throughout the Galaxy.

First of all, the SRa and SRb stars and the irregulars with C or S spectra
have the same distribution as the long-period variables with the same spectra,
that is, they form a flat system, a typical Population I. SRc stars also form a
very flat system with nearly all of them being located less than 100 parsecs
from the galactic plane.

The distribution of SRa and SRb stars with M spectra is similar to that of the long-period stars; in other words, they are included either in an inter-mediate Population I or in a Population II.

The semi-regulars and irregulars of small amplitude, as was noted by Eggen in 1969, belong to the halo population, just as do the Mira stars of period less than 200 days. The RV Tauri stars, whose resemblance to CW stars has already been pointed out, and the SRd stars, also belong to the halo population.

Chapter 5

Miscellaneous Pulsating Variables

The stars we are going to study in this chapter are not a homogeneous group, but really form several different groups with none of them being very numerous. Their amplitudes of variation are often small, so that observing them requires considerable equipment. They are therefore mainly of interest to observers equipped to make photoelectric photometric measurements. It is, however, useful to know about these groups since they all present important problems to theoreticians.

Two of the groups, the dwarf Cepheids and the δ Scuti stars, are located in the instability strip of the H–R diagram that we have already mentioned. We saw in Figure 3 that they form an extension of the Cepheid-RR Lyrae sequence.

Stars of the β Canis Majoris type occupy a well defined position in the H–R diagram, but this is not the case with the α_2 Canum Venaticorum type stars, whose spectra show some remarkable features.

We shall say only a little about Vega and stars resembling it. This is a type not yet ratified, but it does nevertheless constitute a special group. Nor shall we say very much about the ZZ Ceti type of star, a group of white dwarfs showing a variation with a very small amplitude, or about a new group that has just been discovered, the B supergiants.

THE DWARF CEPHEIDS

The small group of dwarf Cepheid stars was for a long time included with the large RR Lyrae family, and it was therefore given the symbol RRs, which it has kept. It was only in 1955 that H. Smith distinguished it as a separate group by noting firstly that the periods are shorter than those of typical RR Lyrae stars, and secondly that they have weaker absolute luminosities.

The typical star is Al Vel (Figure 38), which varies in visual magnitude between 6.41 and 7.13. Its normal period is 0.1116 day or 2 h 40 min, but in fact there are several superposed periods that make the light curve very complex. A modulation period P_b of 0.3792 day has been shown to exist.

77

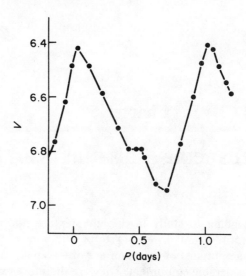

Figure 38. Light curve of Al Vela (type RRs)

About 60 of these stars are known and Table 21 gives the properties of 8 of them. The periods range from 0.05 to 0.25 day, the most frequent period being between 0.10 and 0.20 day.

Table 21. Some dwarf Cepheids

Designation	V	ΔV	P days	P h	min	P_b (days)	Spectra
SX Phe	6.78	0.73	0.0549	1	19	0.1923	A2 V
CY Aqr	10.42	0.73	0.0610	1	28	0.1777	A2–A7
ZZ Mic	9.27	0.35	0.0671	1	37		A3–A8
DY Peg	10.00	0.56	0.0729	1	45	0.2554	A3–A9
Al Vel	6.41	0.72	0.1116	2	40	0.3792	A2–F2p
V702 Sco	7.60	0.67	0.1152	2	46	0.4973	F0
DY Her	10.13	0.49	0.1486	3	34		A7–F4III
VZ Cnc	7.53	0.72	0.1784	4	17	0.7163	A2–F2III

BL Cam is an interesting case in this group. This star was initially suspected of being a white dwarf from its colour index. Its variability and its type (RRs) were announced by Berg and Duthie in 1876. The period is extremely short at 0.0391 day (56 min) for an amplitude of 0.35 magnitude. Its absolute magnitude, measured by using its proper motion, is $+5.5$, but its spectrum is type B, which places it among the sub-dwarfs between the main sequence and the white dwarfs. If the absolute magnitude is accurate, BL Cam is a special case.

Another star worth looking at is one of the best observed of this type: CY Aqr. Superposed on its very short principal period P_1 of 0.061038 day (1 h 28 min) is an even shorter secondary period P_2 of 0.04548 day. The modulation period P_b is 0.17766 day. In this connection, Table 21 shows that in all cases the modulation period is 3–4 times greater than the primary period. We have already seen that this is not so with RR Lyrae stars, where it is more than 50 times greater, and we shall also see that it is not the case with δ Scuti type stars either.

CY Aqr has an appreciable amplitude of about 0.75 magnitude. What is remarkable about this star is that the growth phase of its brightness cycle only lasts about 20 min. The rate of increase to the maximum is thus about 2.2 magnitudes per hour, much greater than that of other pulsating variables and of the same order as that of many eruptive variables!

RRs stars are of the moderately late spectral types A or F with appreciable variations between the maxima and minima. Variations in radial velocity are also large (amplitude 100 km/s). It should be noted that the spectral changes have the same period as the photometric variations and are thus closely connected with the pulsation.

It has long been accepted that the dwarf Cepheids belong, like the RR Lyrae stars, to Population II. However, some indications such as the relative abundances of the elements in the spectra now incline us to think that they belong to a much younger population.

These stars have a period–luminosity relationship all their own: the absolute magnitudes range from $+1$ for the longest periods to $+4$ or $+5$ for the shortest.

δ SCUTI STARS

These stars were initially confused with the dwarf Cepheids, which are of similar spectral types. They are distinguished, however, by a much smaller amplitude, often less than a tenth of a magnitude. The light curves of one of them, β Cas, are shown in Figure 39.

After a lot of systematic searching, carried out mainly in the years 1970–75, we now know more than 100 of these stars. Table 22 gives the properties of 8 of the best observed of them.

The periods are even a little shorter than those of the dwarf Cepheids, ranging from 0.02 to 0.25 days, but there does not appear to be a very marked modal value. We can only say that a good proportion of the periods are within the range 0.05–0.15 day.

Superposed periods can also be detected in this type of star. In some cases a modulation period has been found, some 20–50 times longer than the principal period. For example, δ Scuti itself has $P_1 = 0.1937$ day and $P_b = 5.2477$ days (or about 27 times the principal period).

Spectral types range from A5 to F5 and appreciable fluctuations are

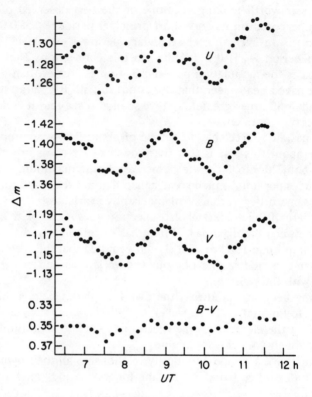

Figure 39:   Light curves in *U, B, V* and *B−V* for β Cas (δ Sct type). The ampli-
tude is given with respect to a comparison star. (*From R. L. Millis*, Inf. Bulletin of
Variable Stars, *No. 137, 1966. Reproduced by permission of the Konkoly Observatory
Budapest*)

Table 22.   Some bright δ Sct type variables

Designation	V	ΔV	P days	P h	min	Spectra
ε Cep	4.17	0.05	0.042	1	00	F0 IV
τ Peg	4.60	0.02	0.0543	1	19	A5 IV
FZ Vel	5.18	0.02	0.065	1	34	F0 III
χ₂ Boo	6.09	0.04	0.0678	1	38	F3 V
τ Cyg	3.73	0.03	0.083	2	00	F2 IV
β Cas	2.78	0.04	0.1043	2	30	F2 III–IV
V474 Mon	6.29	0.07	0.1352	3	14	F2 IV
δ Sct	4.71	0.15	0.1937	4	40	F3 IIIp

observed in the radial velocity, always with a period equal to the photometric period. The stars are slightly less bright than the dwarf Cepheids with absolute magnitudes from +2 to +5.

The δ Scuti stars are young and belong to Population I. They are often present in relatively young galactic clusters: several of them are known in the Hyades cluster and 10 or so have been discovered in the Praesepe cluster (NGC 2632).

Many astrophysicists think that, in spite of several photometric differences between them, particularly the amplitudes, dwarf Cepheids and δ Scuti stars can be grouped into a single family. It should be noted, however, that there are many more δ Scuti stars near the sun than there are dwarf Cepheids: limiting the count to those with apparent magnitudes greater than 6.5, there are more than 50 δ Scuti and only 3 or 4 dwarf Cepheids. It should also be pointed out that many δ Scuti stars are spectroscopic binaries with periods that may be short (as in KW Aur with $P = 3.789$ days), but may also be long (FM Vir, 38.324 days). There is undoubtedly a relationship between the binary nature of these stars and their amplitudes.

β CANIS MAJORIS STARS

Before describing this small group, it should be pointed out that it was originally called the 'β Cepheid type' since it was β Cephei that was the first to be discovered. E. Frost recognized variations in its radial velocity in 1901 and in 1906 determined the period from them. P. Guthnick measured the photometric variations in 1913.

The General Catalogue of Variable Stars prefers to use the designation β Canis Majoris because this star is the brightest of the group, but it should be noted that this decision is not always respected and many authors continue to use the older name.

These stars also have short periods, with a modal value between 0.15 and 0.20 day, although one exception should be noted: ES Vul has a period of 0.610 day.

The amplitude of variation in brightness is always very small, generally being less than 0.1 magnitude, while the radial velocity varies with the same period as the brightness but with an appreciable range, often more than 100 km/s (see Figure 40). These latter variations are also complex, often out of phase with the light curve. The theory of the effect has not been fully worked out and does not explain all the observed peculiarities.

More than half these stars have two pulsation periods, but these differ very little from each other. Thus, the typical star β CMa has $P_1 = 0.25002246$ day and $P_2 = 0.25130003$ day, so that the modulation period is particularly long: $P_b = 49.1236$ days, which is nearly 200 times the principal periods. It is smaller for other stars such as σ Sco, for which $P_b = 8.252$ days, or DD Lac, for which it is 8.5760 days.

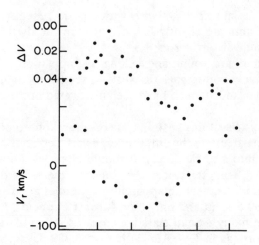

Figure 40. Light curve and radial velocity of σ Scorpii (β CMa type). (From L. Rosino, *Le Stella Variabili*, Caelum, 1980)

Several stars have shown slight variations in period. σ Sco, for instance, changed from a period of 0.246829 day in 1916 to 0.246846 day in 1954 and then decreased to 0.2468318 in 1965. Such variations are rare, however, and most of the periods have been constant over more than 50 years.

More than 50 stars of this type are known and Table 23 gives details of 8 of them. They are sub-giants or giants, but what is most striking about them is the remarkable uniformity of their spectral types, which range only from B0 to B3 (see Figure 41). Their absolute magnitudes lie between −3 and −5 and Table 23 shows that these too obey a period–luminosity relationship that is peculiar to them, the longer periods corresponding to the brightest stars.

Table 23. Some bright β Canis Majoris variables

Designation	V	ΔV	P days	P h	min	Spectra	M_V
δ Cet	4.06	0.06	0.161	3	52	B2 IV	−3.3
δ Lup	3.22	0.03	0.165	3	58	B1.5 IV	−3.4
β Cep	5.22	0.04	0.1905	4	34	B2 III	−4.1
DD Lac	5.22	0.18	0.1931	4	38	B1.5 III	−4.1
ξ₁ CMa	4.35	0.06	0.2096	5	02	B1 III	−4.2
σ Sco	2.89	0.12	0.2468	5	53	B2 IV	−4.3
β CMa	1.98	0.07	0.2500	6	00	B1 II–III	−4.7
V986 Oph	6.12	0.09	0.291	6	59	B0 IIIp	−4.8

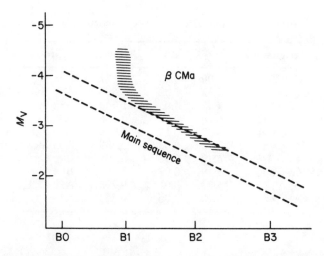

Figure 41. Position of β Canis Majoris stars in the H–R diagram

These are young stars found in O–B stellar associations,* especially in Per I, and in young galactic clusters. Several of them are spectroscopic binaries, and these include β Cep (P = 10.893 days), σ Sco (34.23 days) and α Virginis (Spica), whose period as a binary is 4.0245 days compared with a P_0 of 0.1738 day.

This type of star is as yet poorly understood theoretically and at present no model has been established that explains the observed phenomena.

α_2 CANUM VENATICORUM STARS

We now look at a rather peculiar stellar group which was only distinguished around 1950. It is characterized by an A-type spectrum (B9–A5) which shows a *completely abnormal* abundance of certain heavy metals and, in contrast, a marked deficiency of elements such as oxygen and light metals such as calcium. From the spectral point of view, three groups can be distinguished: first, those which show mainly silicon (Si) lines; secondly, those in which manganese (Mn) lines are predominant; and finally, stars showing lines of chromium (Cr), strontium (Sr) and rare earths such as europium (Eu). It can be seen from Table 24 that the spectral types are denoted by Ap followed by the most abundant elements (for example, A0p Cr, Eu for CS Vir).

These stars are at the same time photometric variables, with an amplitude that is always small and a period that is often long, spectroscopic variables, with time variations in line intensities and profiles, and magnetic variables, with magnetic fields that are strong (several thousand gauss) but that also vary greatly, as can be seen in Figure 42 relating to α_2 CVn.

* Stellar associations are defined in Chapter 5 of Part 3.

Table 24. Some α_2 Canum Venaticorum variables

Designation	V	ΔV	P (days)	Spectra
CU Vir	5.00	0.03	0.521	B9p Si
χ Ser	5.33	0.02	1.596	A0p Eu
ι Cas	4.52	0.03	1.740	A5p Sr.Cr
GO And	6.14	0.04	2.156	A0p Cr.Eu
CG And	6.37	0.10	3.740	A0p Si.Eu
α_2 CVn	2.92	0.03	5.439	A0p Si
V451 Her	6.30	0.08	6.007	B9p Cr.Eu
GS Tau	5.19	0.07	7.229	B9p Si
CS Vir	5.73	0.04	9.295	A0p Cr.Eu
AF Dra	5.18	0.06	20.273	A3 IIIp

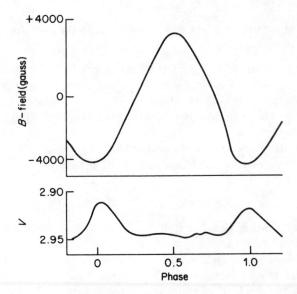

Figure 42. Magnetic and photometric variations in α_2 CVn

It is important to note that the periods of all these variations, photometric, spectroscopic and magnetic, are the same. This is interpreted as showing that the observed period is that of the rotation of the star about its own axis.

Several hundred stars of type Ap have been catalogued and of these there are now over 100 with known variability. Table 24 gives the properties of 10 of the brightest of them.

The periods vary widely. There is undoubtedly a small peak in the frequency of occurrence around 2–3 days, but they extend over the range between 0.5 day and several hundred days (314 days in the case of γ Equ).

The visual amplitude is small, at most 0.1 magnitude. Measurements made at several colours show that this amplitude varies enormously with wavelength: for UU Com, it goes from 0.02 magnitude in V to 0.80 in U. In addition, there is a curious effect observed that has not so far been explained: for each star, there is a wavelength at which there is *no variation in brightness*. Finally, if the brightness is a maximum in the ultraviolet, it is a minimum in the red, and vice versa. The short and long wavelengths are out of phase!

Strictly, these stars should not be considered as pulsating variables since the observed period is none other than the time of rotation of the star about its own axis. However, in the last few years pulsations of very short periods have been discovered: 30 min for UU Com and just over 6 min for DO Eri. These are pulsations of very small amplitude and are independent of the 'normal' photometric variations in the stars.

How can the normal variations be explained? It is thought that the over-abundant metals in the atmosphere are not uniformly distributed but, trapped by the magnetic field, are localized at certain points. This leads to an irregularity in the brightness made perceptible by the rotation. Work carried out in recent years has shown that the localization takes the form of rings or segments and not just of irregular patches.

We may add that some of these stars are known to be spectroscopic binaries. The periods as binaries are different from the periods of the light variation; in other words, the period of revolution of the pair is different from that of the star's rotation.

During the last 10 years, this group has been the subject of much research, both theoretical and observational, since the anomalies encountered in them are of great interest, particularly for studying the structure of internal layers of stars.

A NEW TYPE?

In 1955, O. Struve reported the possible existence of a new group of variables that he called 'Maia' after the star he first studied (20 Tauri, one of the Pleiades). The characteristic of the group is a very rapid but not very regular light variation associated with variations in radial velocity.

In this group must be included the brightest star in the northern sky, Vega (α Lyrae). Variations in its radial velocity were detected from 1888 by H. Vogel and its variability was observed photoelectrically in 1915 by P. Guthnick, who continued his observations until 1930. He reported that the amplitude varied from 0.03 to 0.08 magnitude, while the radial velocity oscillated between -0.9 and -17.8 km/s; the photometric and spectroscopic variations were opposite in phase (Figure 43). It is interesting that, in spite of the observations of Guthnick and those of Belopolski, who confirmed the variability, the International Astronomical Union had, in 1928, included Vega among the list of stars to be used as comparison stars for the determination of radial velocities! This must have falsified numerous results.

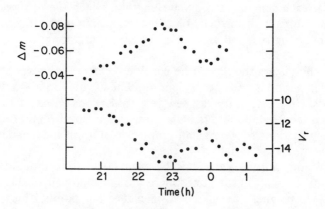

Figure 43. Photometric and spectral variations of Vega

Recent observations have confirmed the spectral and photometric variations in Vega. By studying the spectrum in the near-infrared, D. Morrison and T. Simon (1973) showed that the line profiles are constantly changing. Belopolski was reported to have found a period of 0.179 day, but this has not been confirmed and it even seems that the variation is not periodic, unless there are several superposed periods.

Apart from Maia and Vega, only a few stars are known, all very bright, which could be included in the same group: γ Umi, γ Gem and θ Vir.

ZZ CETI STARS

This is another special group. The typical star, ZZ Ceti, is a white dwarf of spectral type DA and visual magnitude 14.09. It shows periodic but not radial oscillations; in other words, the variations are not due to changes in the radius. They arise instead from shock waves which traverse its atmosphere. For ZZ Ceti, there are two periods of 213 and 274 s.

This group numbers about 20 stars at present, all white dwarfs with DA spectra and absolute magnitude between +11 and +13. Their periods range from 2 to 20 min and the amplitude of the variations is always small, 0.2–0.3 magnitude at most.

These stars are of great interest to astrophysicists because a study of their pulsations gives a better understanding of the atmospheres of white dwarfs. It turns out that the analysis uses a method similar to that in seismology, which gives us information about the earth's crust. However, they are of little interest to amateurs, since they are all very faint (the brightest is

ZZ Psc with $V = 13.75$), have a small amplitude and need large instruments and sophisticated techniques for their observation. The oscillations often have superposed on them (CY Leo or GP Com, for example) rapid eruptive variations (the 'flickering' that we shall be talking about in connection with eruptive variables). The variations in these stars are thus complex and difficult to analyse.

VARIABLE B SUPERGIANTS

The discoveries are never ending: yet another new type of pulsating variable can be added to those already described. These are hot supergiants of spectral type A or B, with absolute magnitudes of -7 or -8 and thus with high luminosity. One of the best observed of these stars is κ Cas. Its maximum magnitude in V is 4.16 and its spectrum is B1 Ia. It varies by 0.04 magnitude in 7 days but it should be pointed out that these variations are not strictly periodic. There are in addition long variations that seem, according to Percy, to be superposed on the semi-regular ones.

Among the 15 stars recognized as belonging to this group is α Cyg with a spectrum A2 Ia, considered as the typical star. According to A. Maeder, the periods are all of the same order (5–10 days) and the amplitude of variation is always less than 0.1 magnitude. These stars are therefore not spectacular variables, but their study is important to an understanding of the internal structure of the hot supergiants.

Part 3

ERUPTIVE AND CATACLYSMIC VARIABLES

Once again, we group together into one large class, that of cataclysmic or eruptive variables, stars that differ greatly from each other but have one feature in common: the variation is an abrupt and explosive event which may occur repeatedly or just once and which may be very powerful or very feeble. The mechanism differs from one stellar type to another and we shall therefore describe this for each type in the relevant chapter. We should add that the term 'cataclysmic variable' is generally used to describe novae and supernovae, and 'eruptive variable' to refer to stars whose explosions are more moderate.

There are six chapters in this section of the book: the first is devoted to novae, the second to supernovae and the third to a wide variety of nova-like groups conveniently classified under the name 'novoids.' The fourth chapter deals with dwarf novae, interesting stars with eruptions occurring in rapid succession. The fifth is confined to variables associated with diffuse nebulae, an important but complex group. Finally, Chapter 6 concerns variable red dwarf stars or flare stars.

Chapter 1

Novae

Of all the types of variable star, novae and supernovae appeal most strikingly to the imagination of observers because of the speed and amplitude of their variations in brightness. Speaking of the impression left by the Perseus nova of 1901, the American astronomer Henry N. Russell said, 'It was difficult to avoid the impression that it made a noise.'

While many novae are faint, some have been extremely bright. Table 25 lists the 23 novae whose brightness exceeded magnitude 6.0 that have appeared since the beginning of this century. Among these, the most spectacular was undoubtedly Nova Aql 1918, catalogued as V603 Aql. It reached a magnitude of -1.1 and was nearly as bright as Sirius.

However, we have chosen to describe another slightly weaker nova which was the last bright one to appear: Nova Cyg 1975, designated as V1500 Cyg. This appeared on the 29 August 1975 and was seen by many observers. The Central Bureau of the IAU at Copenhagen received several dozen telegrams announcing the discovery! Since it is conventional to associate the name of the discoverer with the nova (see Table 25), the Union decided that in this case the name would be that of the sender of the first telegram, the Japanese T. Osada.

When first observed, the nova had a magnitude of 3, but on photographic plates taken only a few hours previously it was found to be of magnitude 8 (see Figure 44). Its increase in brightness was therefore very rapid. The following day, 30 August, it reached its maximum magnitude of 1.8 and then started its decline immediately. At first this was rapid also, interrupted only by short-period oscillations. By 4 September, 5 days after its maximum, it had fallen to magnitude 5, losing nearly one magnitude per day. The decline slowed a little: by 15 September it was of magnitude 7.5, by the end of September of magnitude 8 and by the end of December of magnitude 10.

The Cygnus nova has enabled considerable progress to be made in the understanding of these stars; there had not been a nova as bright as this for more than 30 years and it was therefore the first to be the subject of a large number of observations made with powerful instruments over the whole spectral range. We shall return later to some of the results obtained.

Table 25.   Novae brighter than magnitude 6 occurring in this century
(excluding recurrent novae)

Designations		Magnitudes				Date of	
Nova	Star	Max.	Min.	Δm	Type	maximum	Discoverer
Per 1901	GK Per	0.2	14.1 V	13.9	Na	24.02.1901	Anderson
Gem 1903	DM Gem	4.8	16.5 V	11.7	Na	06.03.1903	Turner
Lac 1910	DI Lac	4.3	14.9 pg	10.6	Na	17.08.1910	Espin
Gem 1912	DN Gem	3.6	15.76 B	12.2	Nb	15.03.1912	Enebo
Mon 1918	GI Mon	5.2	15.1 pg	9.9	Na	01.01.1918	Wolf
Aql 1918	V603 Aql	−1.1	11.97 B	13.1	Na	11.06.1918	Bower
Cyg 1920	V476 Cyg	2.0	17.1 pv	15.1	Na	25.08.1920	Dennings
Pic 1925	RR Pic	1.2	12.8 pg	11.6	Nb	09.06.1925	Watson
Aql 1927	EL Aql	5.5	19.0 pg	13.5	Na	12.06.1927	Wolf
Tau 1927	XX Tau	6.0	16.5 pg	10.5	Na	01.10.1927	Schwassmann, Wachmann
Her 1934	DQ Her	1.3	15.6 pg	14.3	Nb	18.12.1934	Prentice
Lac 1936	CP Lac	2.1	15.6 pg	13.5	Na	21.06.1936	Gomi
Aql 1936	V368 Aql	5.0	15.7 pv	10.7	Na	22.09.1936	Tamm
Sgr 1936	V630 Sgr	4.0	14.4 pg	10.4	Na	04.10.1936	Okabayasi
Mon 1939	BT Mon	4.5	16.8 pg	12.3	Na	09.09.1939	Wachmann
Pup 1942	CP Pup	0.5	<18 pg	>17.5	Na	12.11.1942	Dawson
Lac 1950	DK Lac	5.0	15.5 pg	10.5	Na	23.01.1950	Bertaud
Her 1960	V436 Her	3.0	17.5 pv	14.5	Na	05.03.1960	Hassel
Her 1963	V533 Her	3.0	14.9 pg	11.9	Na	31.03.1963	Dahlgren
Del 1967	HR Del	3.70	12.38 B	8.68	Nb	14.12.1967	Alcock
Vul 1968	LV Vul	5.17	16.9 B	11.7	Na	18.04.1968	Alcock
Ser 1970	FH Ser	4.5	16.1 pg	11.6	Na	16.02.1970	Honda
Cyg 1975	V1500 Cyg	1.8	<21 B	>19.2	Na	31.08.1975	Osada

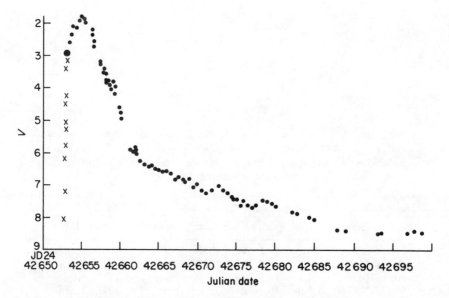

Figure 44. Light curve for Nova Cyg 1975. The circle indicates its discovery, the dots indicate photometric measurements and the crosses refer to photographic records made before the discovery and examined later

Another important point should be made: the pre-nova does not appear in the Mount Palomar Sky Survey which records objects up to photographic magnitude 21. The amplitude of the variation is thus greater than 19 magnitudes; the ratio between the maximum and minimum brightness is therefore greater than 40 million! This is a record for a nova. Its absolute magnitude, determined by G. de Vaucouleurs, was −10.3 at the maximum, making it in fact one of the most powerful known.

For the moment we leave this interesting star to point out that novae have two designations: one as a nova indicating the year of its appearance (for example, Nova Per 1901), the other as a variable (GK Per).

There are three large classes of novae: fast novae (denoted by Na in the catalogue), in which the decline starts immediately after the maximum; slow novae (Nb), which remain a long time around their maxima; and recurrent novae (Nr), which are stars of small amplitude but with nova-like outbursts that recur again and again.

We now examine each of these types in more detail.

FAST NOVAE

These stars are characterized by an extremely abrupt rise to their maxima, some increasing by more than 10 magnitudes in one day. The very speed of the rise means that it is not often observed and even then only in its upper part. The decline is also fairly rapid. An arbitrary but practical criterion to characterize them has been created: this is based on the time T_3 for the nova to lose 3 magnitudes with respect to the maximum. For Nova Aql 1918 T_3 = 8 days, for Nova Per 1901 17 days and for Nova Pup 1942 12 days. One of the fastest, as we have said, was Nova Cyg 1975, for which T_3 = 5 days.

Figure 45 shows the light curves of three fast novae: GK Per, V603 Aql and CP Lac. Oscillations have shown up in some novae during their decline, a phenomenon that was particularly evident in Nova Per 1901; its oscillations lasted for 3 months and had an amplitude often exceeding one magnitude. On the other hand, others such as Nova Lac 1936 or Nova Cyg 1975 have shown only small fluctuations.

There have been a large number of fast novae: out of 121 galactic novae whose type has been determined, 82 belong to type Na.

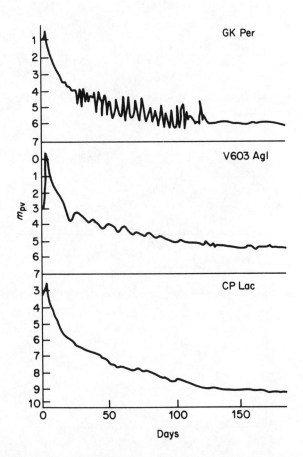

Figure 45.   Light curves extending over 6 months for three fast novae: GK Per (N. 1901), V603 Aql (N. 1918) and CP Lac (N. 1936). (*From Ch. Bertaud,* Annales de l'Obs. de Paris, **9**, *No. 1, 1945. Reproduced by permission of l'Observatoire de Paris*)

SLOW NOVAE

The nova Del 1967 (HR Del) is a good example of this type. The star passed from a magnitude of 12 to one of 6 in 1 month and then continued to increase slowly with the maximum magnitude of 3.7 being reached 5 months later. The decline was even slower, as can. be seen from Figure 46.

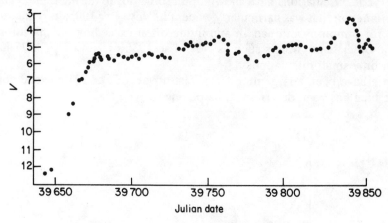

Figure 46. Light curve for Nova Del 1967 (HR Del)

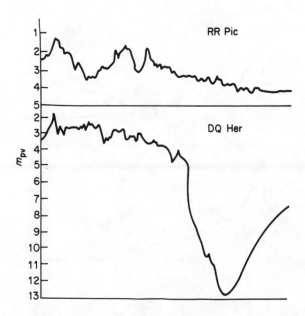

Figure 47. Light curves extending over 6 months for two slow novae: RR Pic (N. 1925) and DQ Her (N. 1934). (From *Ch. Bertaud*, Annales de l'Obs. de Paris, **9**, *No. 1, 1945. Reproduced by permission of l'Observatoire de Paris*)

The light curve of Nova Her 1934 (DQ Her) had a different shape (Figure 47). Discovered on 13 December 1934 by Prentice at a magnitude of 3.4, it reached its maximum magnitude of 1.5 9 days later. It declined until April 1935 (T_3 = 94 days), and then abruptly lost 8 magnitudes in less than 1 month. Following that, there was a new but weaker maximum of magnitude 6.8 in September, with a subsequent very slow decrease.

Nova Pic 1925 (RR Pic) was completely different again: it had three maxima on the 8 June, 28 July and 10 August 1925, and its very slow decline was marked by oscillations.

These stars are less numerous than fast novae, only 32 being known out of the 121 known novae of all types.

I have considered the ultra-slow novae, such as RT Serpentis, or the objects of the η Carinae type to be completely distinct from the 'classical' slow novae. We shall deal with these in the chapter on novoids.

RECURRENT NOVAE

Recurrent novae are stars that show repeated nova-like variations. Table 26 gives details of a few that are certain.* The brightest of them has been

Table 26.   Recurrent novae

Designation	Dates of maxima	Maximum m_V	Δm_V	Intervals (days)
T CrB	12.08.1866	2.0	8.6	
	02.02.1946	1.8	8.8	29126
V616 Mon	08.11.1917	11.5	8.5	
	21.08.1975	11.3	8.7	21 106
RS Oph	30.06.1898	4.5	7.8	
	12.05.1933	4.6	7.7	12 734
	31.08.1958	4.8	7.5	8 115
	21.10.1967	5.3	7.0	3 471
	26.01.1985	5.4	6.9	6 308
T Pyx	28.05.1890	7.5	7.0	
	02.05.1902	7.2	7.3	4 353
	06.04.1920	6.6	7.9	6 549
	21.11.1944	7.0	7.5	8 995
	13.01.1967	7.0	7.4	8 086
U Sco	20.05.1863	9.0	10.0	
	12.05.1906	8.8	10.2	15 695
	23.06.1936	8.8	10.2	11 999
	25.06.1979	8.9	10.1	15 708

---

* We shall deal with V616 Mon in the chapter on X-ray transients (Part 5, Chapter 2).

T CrB, which had two maxima of magnitude 2, one in 1866 and the other in 1946. RS Oph has had four maxima separated by intervals of 35, 25 and only 9 years. U Sco has also had four maxima but at longer intervals of 43, 30 and 43 years. The most active has been T Pyx, whose five maxima were separated by 12, 18, 24 and 23 years.

Some of these novae have been very fast (T CrB with $T_3 = 6$ days or RS Oph with $T_3 = 9$ days), while others such as T Pyx have light curves resembling those of slow novae.

It is important to point out that the maxima of a given star are the same as regards both the amplitude and the shape of the light curve. This clearly shows, as we have seen, that the occurrence of the nova, in spite of its violence, does not appreciably alter the structure of the star.

WZ Sagittae is a special case: it was long considered to be a recurrent nova, with three explosions occurring in November 1913 (magnitude 8.5), June 1946 (8.7) and December 1978 (8.7). However, spectroscopic observations made particularly by Rosino during its last maximum showed that it is not to be classified with the recurrent novae but with the dwarf novae; its spectrum is similar to that of a U Geminorum star and the speeds of ejection of the gases are much lower than is the case for recurrent novae. In addition, it exhibits a spectroscopic oscillation that we shall mention later and that is a feature of dwarf novae. Finally, its absolute magnitude of approximately +10 is different from that of the novae.

THE SPECTRA OF NOVAE

The spectra of these stars is so distinctive that they do not fit in to the usual classification and a special category, type q, has been created to designate them.

Once again, we take the Nova Cyg 1975 as an example since it has been very closely observed spectroscopically, particularly by A. Woszczyk and his colleagues. The first spectra were obtained on 29 August 1975 shortly after its discovery. They had an intense continuous background and the H and K lines of ionized calcium in the absorption spectrum, showing radial velocities of the order of 1000 km/s.

From the maximum, emission lines appeared: those of the Balmer series of hydrogen, and of ionized calcium and iron. From 5 September ionized helium appeared, but the emission lines from ionized calcium disappeared after 7 September, and those of the iron several days later. At the same time, the line profiles changed and the radial velocities increased: the velocities from the absorption lines increased to 2600 km/s from 4 September (see Figure 48) and those from the emission lines were even higher (3000–3400 km/s).

From 8 September, the 'nebular' spectrum appeared, so called because it resembled that of gaseous nebulae. It showed lines belonging to highly

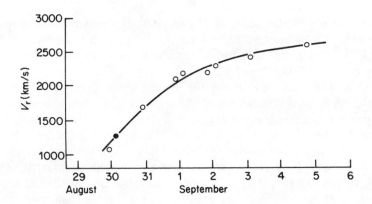

Figure 48. Variation of the radial velocity of Nova Cyg 1975. (*From A. Woszczyk, S. Krawczyk and A. Strobel,* Inf. Bulletin of Variable Stars, *No. 1072, 1975. Reproduced by permission of the Konkoly Observatory Budapest*)

ionized elements (oxygen, helium, neon, carbon, etc.). Following that, the emission lines broadened considerably and several months after the explosion they formed very wide bands.

Other novae have shown spectral fluctuations of the same kind. We certainly cannot examine these in detail, but merely indicate that in general the 'nebular' spectrum only appears when the nova has already diminished by several magnitudes.

One point seems surprising: some novae have shown appreciable radiation in the infrared. This is the case for FH Ser (Nova 1970), studied in the far-infrared (2–25 μm) by Hyland and Neugebauer and by Geisl, Kleinman and Low. A little after its visual maximum, it became one of the brightest infrared stars in the sky. The recurrent nova RS Oph has shown the same effect, which is still not very well explained. Is there a red component present which is activated by the explosion, or is it something else? We still do not know.

SECONDARY PHENOMENA

Nova Persei has displayed a curious effect: in August 1901, 6 months after the maximum, photographs showed the presence of a large halo extending to 9″ from the nova. In October it extended to 11″, in February 1902 to 17″ and in November 1902 to 24″. Such growth implied an enormous velocity of expansion! In November 1902, Perrine obtained the spectrum of this halo and to everybody's surprise it was the spectrum of the nova at its maximum and not the nebular spectrum that the nova showed at that time. The explanation was later given by Kapteyn: the halo is the *image* of the explosion, which is reflected from a dark cloud surrounding the star. The diameter of this image thus increases at the speed of light. This is indeed what is deduced by knowing the distance between the nova and ourselves.

Independently of this halo, Nova Per 1901 showed a real nebulosity discovered in 1916 by Barnard and with a diameter which had grown to 38″ by 1934. This corresponds to a mean increase of more than 1000 km/s. After that, it gradually disappeared.

Other novae have also shown the same phenomenon of nebulosity in which its appearance coincided with the appearance of the nebular spectrum. Examples are the novae Aql 1918, Oph 1919, Cyg 1920 and Pic 1925.

Several novae display another effect, that of an apparent doubling. In 1928, Van den Bos in observing Nova Pic 1925 saw it double and then treble a few days later. This star was then closely followed and it was seen that, although the angular position of the components did not change, the distance apart grew by 0.2″ per year. The nova Her 1934 showed the same effect, discovered by Kuiper: the angular position (131°) remained the same, but the distance of the components changed from 0.21″ in July 1935 to 0.62″ in March 1937 and to 1″ in August 1939. This is not in fact a true doubling of the star, but an ejection of enormous gaseous masses with a stellar appearance. After several years, these masses lose this appearance and dissipate.

AMPLITUDE AND ABSOLUTE MAGNITUDE

Novae have varied amplitudes that range from 7 to more than 19 magnitudes, but the value cannot always be determined since the star is often very faint at its minimum. Nevertheless, the amplitudes are known for 76 stars and Figure 49 shows their distribution. There are two peaks in the frequencies with which they occur, one for an amplitude of 9 magnitudes and the other for one of 12. However, it is probable that large amplitudes are more common but are not known.

Recurrent novae have very small amplitudes ranging from 8 to 10 magnitudes. Slow novae, such as DQ Her, can have large amplitudes, but examination of Figure 49 shows that novae with amplitudes greater than 13 magnitudes are mostly fast novae. Note that this figure does not include all novae, and note also that it excludes the greatest known amplitude, that of Nova Cyg 1975, which is more than 19 magnitudes. This star is considered by many authors to be an exception—intermediate between a nova and a supernova.

A lot of work has been done towards the establishment of absolute magnitudes. Standard methods for determining distances (parallax, proper motion) can hardly be used here, given the distance of novae (from a few hundred to several thousand parsecs). Other methods have therefore been used: firstly, by measuring the intensity of interstellar lines (intensity increasing as the distance of the object increases) and secondly by obtaining the apparent velocity of expansion of the nebulosity (enabling the distance to be known if the radial velocity of the gases has been determined).

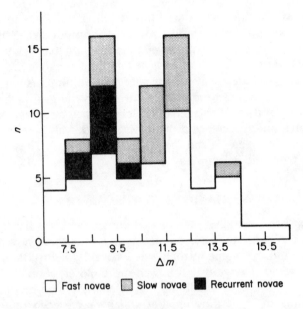

Figure 49. Distribution of nova amplitudes

Table 27 gives several absolute magnitudes both of the maxima and the minima. The minimum values show that there are two groups: first, the 'normal' novae, with minimum absolute magnitudes of $+3$ to $+5$; and secondly, a small group of faint stars, such as DQ Her, CP Pup or V1500 Cyg, whose absolute magnitudes are approximately $+8$.

Table 27. Absolute magnitudes of several novae

Designation	V_{max}	V_{min}	ΔV	Mv_{max}	Mv_{min}
Na DK Lac	5.2	15.1	10.5	− 7.8	+2.7
DI Lac	4.5	15.1	10.6	− 6.5	+4.1
V603 Aql	−1.1	11.6	12.7	− 9.0	+3.7
CP Lac	2.5	15.8	13.3	− 9.0	+4.3
GK Per	0.2	14.1	13.9	− 9.0	+4.9
V476 Cyg	2.0	17.1	15.1	− 8.9	+6.2
CP Pup	0.5	18.0	17.5	−10.0	+7.5
V1500 Cyg	1.7	<20.5	>19	−10.3	> +8.5
Nb T Aur	4.4	16.0	10.6	− 6.1	+4.5
RR Pic	1 5	12.3	10.8	− 7.1	+3.7
DN Gem	4.0	15.9	11.9	− 7.0	+4.9
DQ Her	1.3	15.1	13.8	− 5.4	+8.4
Nr RS Oph	4.5	12.3	7.8	− 6.1	+1.7
T CrB	1.8	10.6	8.8	− 7.8	+1.0

Two groups can also be detected by considering the maxima, one with absolute magnitudes around −6 and the other around −9, and these correspond to the two groups in the distribution of amplitudes. This therefore shows that there is a correlation between the absolute magnitude at the maximum, the speed of decline (T_3), and the amplitude: the very fast novae and those of large amplitude are also those with greatest luminosity at maximum.

All these results are corroborated by the observation of novae in the Andromeda galaxy and in the Magellanic Clouds. They also have two peaks in their frequencies of occurrence around −6 and −9, and there is also a correlation observed between T_3 and the absolute magnitude. They are therefore no different from the galactic novae.

VARIATIONS AT THE MINIMA

The prenovae are generally not very well known, which is hardly surprising since it is not possible to predict which stars will become novae. However, it has been found that some of them were variables before their explosion.

The postnovae are obviously followed more closely. Some of them have fluctuations that are occasionally appreciable, with some sort of small secondary maxima of short duration but which may exceed one magnitude. This is the case with Nova Per 1901, which varies by 1.5 magnitude, and with Nova Lac 1936 or the recurrent novae T CrB and RS Oph.

High-precision photometry has revealed another type of variation that we shall find in many eruptive variables: this is 'flickering,' which consists of small rapidly varying flares following each other without interruption.

Finally, we point out the case of RS Oph. Tempesti has shown that there is a semi-regular variation (period 70 days) of amplitude 0.6 magnitude. This confirms that there is an M giant in the system linked to a blue star.

DOUBLING

M. F. Walker showed in 1954 that DQ Herculis (Nova 1934) is an eclipsing binary with a very short period of 4 h 39 min. Since then, all the novae bright enough to be usefully observed have been shown to be double. In this case, therefore, doubling is a general feature.

These binaries are formed from a red star that is large but not very massive and a blue star of high density, which resembles a white dwarf. These dissimilar pairs are generally closely bound and have a very short orbital period, as can be seen from Table 28.

There are several exceptions to this structure, such as the recurrent novae RS Oph and T CrB, whose red component is a giant, whereas in the other cases it is a sub-giant. The rotation period is therefore much longer.

In some of these binaries small changes in period which arise from variations in the two stars have been detected. The most interesting case is that of V1500 Cyg. Semeniuk and his colleagues have shown that the period

Table 28. Binary novae. The amplitude Δm is given for the eclipsing binaries; the notation SB implies a spectroscopic binary

Designation	Period			Δm	Nova type
	days	h	min		
V1500 Cyg	0.1384	3	20	0.3 V	Na
V603 Aql	0.1450	3	29	0.36 V	Na
RR Pic	0.1450	3	29	0.23 V	Nb
HR Del	0.1706	4	06	0.4 V	Nb
DQ Her	0.1936	4	39	0.32 V	Nb
T Aur	0.2044	4	54	0.10 V	Nb
T CrB	227.6			SB	Nr

has changed from 0.1410 day at the beginning of September 1975 (the time of the explosion) to 0.1399 day at the end of October and to 0.1384 day in May–June 1976. The period of the binary system may thus have been changed by the violence of the explosion.

THE CAUSE OF NOVAE

The mechanism causing a star to become a nova can be broadly described as follows. Consider a pair of stars such as that shown schematically in Figure 50: one of them a large reddish star of low density which has reached or exceeded its Roche lobe, and the other a very dense white dwarf. Some of the material of the red star escapes and is transferred to the white dwarf. It has been known for several years in fact that the transfer takes place in two stages: the material first falls on to a kind of disc or ring which surrounds the white dwarf and which is called the accretion disc. At a later time, the matter which is attracted by the strong gravitational force from the white dwarf leaves the disc and falls on to the dwarf at high speeds and with great turbulence.

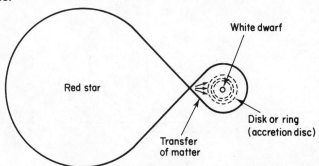

Figure 50.   Processes leading to the formation of a nova

The arrival of the gas at the accretion disc, already hot and becoming hotter still as it falls on to the white dwarf, provokes the eruption: a powerful nuclear explosion is produced in the atmosphere of the star, and the white dwarf suddenly ejects the blanket of 'foreign' matter which covers it: this is the beginning of the phenomenon known as a nova.

The mass of the ejected gas is relatively small: 10^{-4} to 10^{-5} of a solar mass. On the scale of the stars involved, this is extremely minute and the structure of the star is not affected by the explosion, yet it is as much as several times or several dozen times the earth's mass.

The effect may occur more than once: we know that the recurrent novae repeat their explosions at intervals of several decades (Table 26). It is thought that all novae must be recurrent, but for those with large amplitudes the recurrence times must be several thousand or tens of thousands of years. The relationship between the amplitude of a nova and its recurrence time is easy to understand: if the explosions occur frequently, the quantity of material accreted by the white dwarf is small and the explosion will be limited. On the other hand, if the time between two explosions is very long, there is considerable accretion of material and this manifests itself as a very powerful explosion.

GALACTIC DISTRIBUTION

It is interesting to see how novae are distributed through the Galaxy. The distribution of longitudes is remarkable: Figure 51 shows that out of 161 certain galactic novae, 74 occur between longitudes 345° and 15°, in other words, at less than 15° from the galactic centre. Novae are thus particularly numerous in the direction of the galactic nucleus.

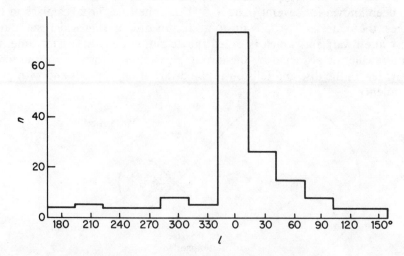

Figure 51.   Distribution of 161 novae among galactic longitudes, ℓ. Note the large concentration around the galactic centre ($\ell = 0°$)

As regards latitude, novae situated in the direction of the centre are never very far from the galactic plane. These are faint novae, located at several thousand parsecs from the sun in the direction of the nucleus and at 400–500 parsecs from the galactic plane. In other directions, several novae are known that are quite a long way from the galactic plane, but very few are more than 1500 parsecs from it.

All this leads us to conclude that novae form an intermediate Population II. However, they do not seem that to form a homogeneous population. Two novae are known in globular clusters that we know form a typical Population II: T Sco in Messier 80 and V1148 Sgr in NGC 6553. In addition, they have been found in three elliptical galaxies (NGC 147, 185 and 205) which also form a typical Population II.

Many novae have been observed in nearby galaxies: in Messier 31 alone about 200 have been detected, which is as many if not more than in our own Galaxy (they are on average of magnitude 16–17). Four novae are known in the Small Magellanic Cloud, six in the Large Magellanic Cloud, and a significant number in other galaxies such as that of the Triangle (M33). In all, about 500 galactic and extra-galactic novae are now known.

Finally, it should be pointed out that the novae observed in other galaxies are of roughly the same absolute magnitude (−6 to −9) and have the same type of light curves as those in our Galaxy. These novae are sometimes used to determine the distance of the galaxies in which they are observed. The results are uncertain, however, since the relationship between the speed of decline and the absolute magnitude is not a rigorous one.

Chapter 2

Supernovae

Supernovae, even more than novae, strike the imagination by the enormous scale of their variations in brightness. Their history begins in 1885: in August of that year, a nova (designated as S And) appeared in the Andromeda spiral Messier 31 and at its maximum reached an apparent magnitude of 5.4—the brightest stars of that galaxy have magnitudes of 15–16. When the distance of M31 was determined by Hubble, it was realized that S And was not a simple nova. The same applied to Z Cen of NGC 5253 in 1895, and to several other stars observed in other galaxies.

A systematic search for such stars, called 'supernovae,' was undertaken in 1926 by Baade and Zwicky, and continued until recent years by Zwicky alone. The search has proved very fruitful: in spite of their comparative rarity, nearly 500 supernovae have been discovered in galaxies, sometimes very distant ones. There have also been searches for supernovae that could have appeared in our own Galaxy and we shall see the results of that work below.

Extragalactic supernovae have a different designation from galactic variables. They are denoted by the year of their explosion, followed by a letter in order of their discovery: SN1970a, SN1970b, etc.

Supernovae exhibit two principal types of light curve, I and II,[*] which are shown in Figure 52. Type I includes the most luminous stars with an estimated absolute magnitude of −19.9 on average. The star loses 3 magnitudes in about 50 days and the decline then becomes slower: 3 further magnitudes in 200 days.

For type II, the absolute magnitude is about −17.8. The decline is at first fairly fast (2 magnitudes in 20 or 25 days), and then a levelling off often occurs, clearly visible in Figure 52, which lasts about 2 months. The star then

---

[*] In 1965, Zwicky separated type II supernovae into types II, III, IV and V according to the shape of their light curves. However, this photometric classification is arbitrary and type V included objects that are not true supernovae (see following chapter).

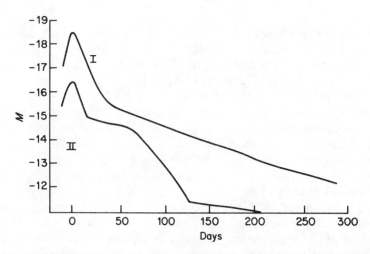

Figure 52. Average light curves for 38 type I supernovae and 13 type II supernovae. (*From R. Barbon, F. Ciatti and L. Rosino,* Astronomy and Astrophysics, **25,**
241, 1973. Reproduced by permission of Springer-Verlag)

loses a further 2 magnitudes in 20–30 days, and after that the decline is
resumed more slowly. We should add that about a third of type II supernovae
do not show the 'levelling off' we have just mentioned.

THE GALACTIC SUPERNOVAE

Supernovae have been observed in our own Galaxy for more than three
centuries. Research has been undertaken to look for stars appearing in historical times which could be considered as supernovae. This is not easy,
since ancient observations were often very inaccurate. In the last 3000 years
at least 20 possible cases have been accepted, and since the beginning of
the Christian era about 15 have been considered altogether, although a few
among them are doubtful. Fortunately, supernovae leave a trace in the form
of a nebulosity or a radio source, and it is these traces which enable us to
confirm that one has indeed occurred.

The first supernova of the Christian era appeared in the year 185 A.D. All
that is known about it is that it was very bright, perhaps of magnitude −2.
Another appeared in 369 A.D. in Cassiopeia, but is still not identified with
certainty.

In 437 A.D., one appeared in Gemini, now assigned a magnitude −3, but
one that appeared in 668 is more doubtful. In contrast, the star that appeared
in Sagittarius in 827 was extraordinary, even though very little observed: its
brightness was comparable to that of the moon: in other words, its magnitude
was apparently approximately −10 at the maximum. So far, unfortunately,
no trace of it has been found.

The first well observed supernova appeared on 4 July 1054 in Taurus.

Astronomers at the time, mainly Chinese and Korean, described it as much brighter than Jupiter. It is estimated that it reached an apparent magnitude of −5 and remained visible throughout the day for 23 days. It was followed (with the naked eye, of course) until April 1055. This supernova, CM Tau, left as its trace the Crab Nebula, to which we shall return later.

Another supernova appeared in 1181 and was undoubtedly very bright, but it was hardly observed. However, Weiler has found an object at its position which is similar to the Crab Nebula and which is at the same time an intense radio source (3C58). Two other supernovae made their appearance in 1203 and 1230, but they were hardly observed at all, and a bright one was observed in China in 1408.

On 11 November 1572, the most famous of the galactic supernovae (B Cas) appeared and was particularly studied by the Dane Tycho Brahe. At maximum, it had the same brightness as Venus and remained visible at midday for several days. Since it occurred shortly after the massacre of Saint-Barthélemy, it was interpreted in France as a sign of divine vengeance! It was followed until April 1574 and then disappeared. Although the astronomical telescope had not then been invented, the observations were numerous enough for a light curve to be plotted. It had the form of a type I supernova (Figure 53). The star left a very faint nebulosity, photographed at Mount Palomar by Van den Bergh, as well as a powerful radio and X-ray source. Another supernova appeared in 1604 in Ophiuchus (V843 Oph). It was discovered by an anonymous Italian, but was observed mainly by Kepler and its light curve was also plotted. At the maximum, it reached a magnitude of −2.5 and was visible for a year. It gave rise to a faint nebulosity discovered by W. Baade in 1941.

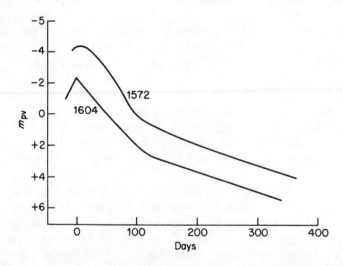

Figure 53. Reconstruction of the light curves of SN 1572 and SN 1604, both of type I

We end the list with a supernova that passed curiously unnoticed, even though it must have been very bright. It left as its trace one of the most powerful known radio sources, Cas A, at the same time as a faint optical nebulosity. The expansion of this nebulosity has enabled us to date its appearance as occurring around 1667 with some degree of uncertainty.

No galactic supernovae have been observed since then, regrettably for astronomers! If one were to appear now, it would, given the means at our disposal, mean a giant leap in our knowledge of these enigmatic stars.

SUPERNOVA REMNANTS (SNR)

The most famous of the supernova remnants is undoubtedly Messier 1, the Crab Nebula, the residue of the star of 1054. It is an object well known to amateurs, since its photograph appears prominently in all astronomy books. It is situated at 1400 parsecs from us and has been shown to be still expanding, with the gas currently escaping at a speed of 1300 km/s. The Nebula is a powerful radio source and also a very intense X-ray source, the strongest among the supernova remnants. The X-ray source is not point-like but diffuse and connected with the Nebula.

Finally, it has a pulsar at its centre, that is, a strongly magnetic neutron star, identified with one of the central stars.* This pulsar, denoted by NP 05-32 possesses one of the shortest periods known for its type: 0.0330945 s or about 33 ms. This very small object turns on its axis 30 times a second!

Table 29.   Features of galactic supernovae. The column headed r gives the distance in parsecs; that headed X-ray source gives the intensity of the source in joules per second

Supernova	Apparent magnitude, m_V	Type	r	Optical traces	Radio source	X-ray source
185 Cen	−8		2500	Faint	PKS1439–62	3×10^{28}
437 Gem	−3	II?	1500	IC 443	Sh 34	2×10^{27}
1006 Lup	−9	I	1300	Faint	PKS1459–41	7×10^{27}
1054 Tau	−5	II	2000	Messier 1	Messier 1	1×10^{30}
1181 Cas	0?	II		Faint	CTA 1	
1203 Sco	−2.5			NGC 4673	CTB 37	
1572 Cas	−4.5	I	5000	Faint	3 C 10	5×10^{29}
1604 Oph	−3	I	9000	Faint	3 C 358	3×10^{28}
1667 Cas	?	I	3400	Faint	Cas A	5×10^{29}

Table 29 summarizes our knowledge of galactic supernovae that are known to have appeared since the beginning of the Christian era. There are, however, nearly 150 remnants of older galactic supernovae known which are in the form of nebulae, radio sources or X-ray sources.

---

* We shall return to the variability of pulsars in Chapter 2 of Part 5.

Some of the nebular remnants, such as the Crab Nebula or Messier 1, are diffuse; others are in the form of a spherical or elliptical ring or loop that may be enormous. This is the case with the Cygnus Loop, of which the well known Veil Nebula is only a part. These loops are often extremely large, 40–50 parsecs in diameter. In some cases, there is no nebulosity since it has been dispersed into space but there is, on the other hand, a radio source that is often very powerful such as that of Cas A, which is shown in Figure 54.

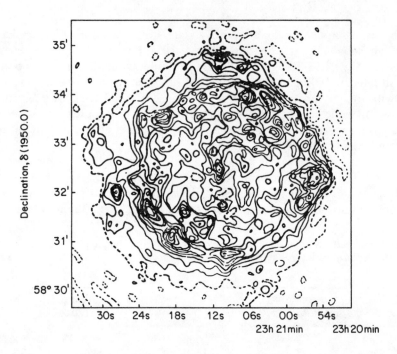

Figure 54.   The radio source Cas A. This source covers a large area, whereas optically only a few faint filaments can be seen. (*From J. Lequeux*, Encyclopédie de l'Univers, *Bureau des Longitudes, Vol. 3, p. 86, 1980. Reproduced by permission of Editions Bordas SA, Paris*)

Weiler has also reported the existence of two types of radio source. With some, the radio map is irregular; this is the case with Cas A (Figure 54) and the maps of the 1572 and 1604 supernovae (Figure 55). In other cases, the map of radio isophotes has an 'oyster shell' structure and this is so with the Crab nebula and the two radio sources G 21.5 − 0.9 and G 74.9 + 1.2 shown in Figure 55.

According to Weiler, the irregular radio structures correspond to type I supernovae: this is known to be so from the light curves of the 1572

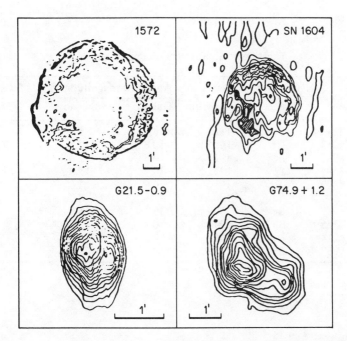

Figure 55. Four radio maps of supernovae, one (G74.9 + 1.2) taken at 21 cm and the others at 6 cm wavelengths. Observations of R. M. Duin, S. F. Gull, K. W. Weiler, A. S. Wilson and P. A. Shaver. (*From K. W. Weiler*, Sky and Telescope, **58**, *414, 1979. Reproduced by permission of Sky Publishing Corp*)

and 1604 supernovae, and no doubt also for those of 1006 and 1667. The concentric structures correspond to type II supernovae, as in the 1054 star and probably in those of 437 and 1181.

EXTRAGALACTIC SUPERNOVAE

Because of their large absolute luminosities, supernovae are visible from a great distance. We have already mentioned that nearly 500 are now known in other galaxies. Table 30 gives the main features of 10 of them, including the designation of each galaxy and its type, the designation of the supernova, the apparent magnitudes of the galaxy and the supernovae, together with the type of the latter. Figure 56 shows the light curves of three well observed supernovae.

It appears that a supernova at its maximum very often radiates as much as the entire galaxy to which it belongs. It can even be brighter if the galaxy is not very large, as in the case of SN 1937c which is more than 5 magnitudes brighter than its galaxy IC 4182.

Supernovae are observed in all types of galaxy, but are particularly numerous in spiral galaxies whether simple or barred. This is illustrated in Table 31,

Table 30. Some extragalactic supernovae

Galaxy	Galactic type	Supernova designation	Apparent magnitude, m_{pg} Galaxy	Supernova	Supernova type
NGC 224	Sb	1885 a	4.3	5.4	I
NGC 5253	Irr	1895 b	10.8	7.5	I
NGC 6946	Sc	1917 a	10.5	12.6	I
NGC 3184	Sc	1921 b	10.7	11.0	I
NGC 4303	SBc	1926 a	10.9	12.6	II
IC 4182	Sc	1937 c	13.8	8.2	I
NGC 6946	Sc	1939 c	10.5	11.8	II
NGC 4725	Sb	1940 b	10.4	12.6	II
NGC 4214	Irr	1954 a	10.3	9.5	I
NGC 5253	Irr	1972 a	10.8	7.8	I

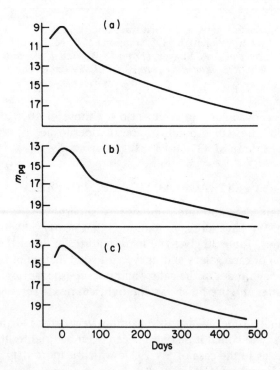

Figure 56. Light curves of three extragalacticsupernovae. (a) for SN 1937c in IC 4182; (b) for SN 1937d in NGC 1003; (c) for SN 1940b in NGC 4725

Table 31. Distribution of supernovae among the
various types of galaxy

Type of galaxy	Number of supernovae
Simple spirals	298
Sb	73
Sc	127
Barred spirals	67
SBc	30
Irregular	31
Elliptical	50
Abnormal	5
Indeterminate type	23

which gives the distribution of 474 of them. And among the spiral galaxies, it is the Sc or SBc (the more open ones) that have the most supernovae. There is thus a relationship between the type of spiral, which is a function of its age, and the number of supernovae.

There is also a relationship between the type of spiral and the absolute magnitude of the supernova at its maximum. According to L. Green, type I supernovae reach an absolute magnitude of -18 in the Sc spirals, -20 in the ellipticals and -21 in the irregulars. Finally, we note that type II supernovae (which belong to Population I) are only found in Sb and Sc spirals, whereas type I supernovae (of Population II) are found in all types of galaxy. This is quite normal since only the Sb and Sc spirals have developed arms possessing a sizeable Population I.

Some galaxies have had two or more supernovae in a few dozen years. For example, NGC 3184 has had 3 in 1921, 1921 and 1937 (two in an 8-month interval); NGC 4321, four in 1901, 1912, 1959 and 1979; NGC 5236, four in 1923, 1950, 1957 and 1968; NGC 5057, three in 1909, 1951 and 1970; and the record belongs to NGC 6946, with five certain and one probable in just over 60 years, 1917, 1939, 1948, 1968, 1969 and 1980.

Many galaxies, on the other hand, some of them very large such as Messier 31, have had only one. This poses the question of how frequently supernovae occur, and we come to this topic below.

THE SPECTRA

The spectra of supernovae are not like any others. They consist entirely of very broad emission or absorption bands, sometimes exceeding 100 Å, whose identification, attempted particularly by R. Minkovski, is very difficult. They involve the spectra of highly ionized elements such as oxygen, hydrogen and nitrogen. One fact is remarkable: the bands change in a regular way with time and always in the same fashion so that, as Maffei points out, an examination of the spectrum made at any time whatsoever enables the date of the maximum to be found.

We shall not enter into a detailed study of these complex spectra, but mention only that there are distinct spectral differences between type I and type II supernovae.

The speeds with which the gases are ejected are enormous, much greater than those for novae. They can be as much as 20 000 km/s for type I and 15 000 km/s for type II.

FREQUENCY OF SUPERNOVAE

It was thought at first that supernovae were very rare (one every two to three centuries), but research into extragalactic supernovae has made astronomers revise their opinion. A few years ago, it was assumed that each galaxy produced on the average one supernova every 50 years, but it is now thought that this frequency must be increased to one every 20–30 years. Some galaxies, as we have seen, have had several supernovae in a very short time.

This would appear to be contradicted by our own Galaxy, in which supernovae seem very rare. In fact, it must not be forgotten that the sun occupies a peripheral position in the Galaxy, so that distant supernovae which explode in the galactic nucleus or in arms located on the other side of the nucleus are practically inaccessible to us, given the amount of absorption produced by dust and gas clouds. The supernovae that have been observed are all within a small region of the Galaxy. G. A. Tamman has been able to deduce from this that the real frequency is much greater than the observed frequency, and reaches at least 4 or 5 supernovae per century. It is thus of the same order as that observed for other galaxies.

The frequency of supernovae is not the same for all galaxies: only one has been seen in Messier 31 over 100 years, and yet several have been observed in much smaller galaxies. We saw just now that there is a relationship between the type of galaxy and the apparent frequency of supernovae; the open Sc and SBc spirals seem more apt to produce them than other types.

THE ORIGIN OF SUPERNOVAE

The amplitudes of supernovae variations are still not very well known, but extragalactic observations have shown that they are visually greater than 20 magnitudes. In addition, measurements made at short wavelengths using satellites have revealed that these stars radiate even more in the far-ultraviolet and X-ray regions than in the visible region. From such information, it is estimated that a supernova releases on average an energy of 10^{44} J, or more than that emitted by the sun over 5000 million years! It is obvious from this that a supernova has a power that is absolutely unparalleled and completely different from a nova, which is several million times less powerful (10^{37}–10^{38} J).

So what is the mechanism behind such a portentous phenomenon? Several theories have been put forward as explanations, but we shall see that none of them is entirely satisfactory.

In the 1960s, W. Fowler and F. Hoyle proposed the following model: when a large star of more than 10 solar masses reaches the end of its evolution, it produces in its central core a significant number of heavy elements, particularly iron. If the temperature at the centre becomes too high (5000–7000 million degrees) because of its continual contraction, the iron atom is destroyed and produces helium nuclei and neutrons. However, unlike the previous nuclear reactions, this process does not produce energy but absorbs it. To provide this energy, the core contracts further so as to release gravitational energy and such a contraction is so abrupt that it is like an implosion and produces a complete change in the stellar structure. The heating is so great that the material surrounding the core explodes, projecting the whole exterior part of the star outwards at an enormous speed. At the same time, the central core, representing 10–20% of its mass, totally collapses. Under the shock, the particles in it (protons, helium nuclei, electrons) fuse to form neutrons, and all that is left is a highly concentrated object, a neutron star or pulsar.

Colgate and White put forward a different theory in 1966: the nuclear reactions which manufacture the heavy elements produce an enormous number of neutrinos, neutral particles like neutrons but with negligible mass, which travel at great speeds through stellar matter. The neutrinos are absorbed by the material and give up their energy to it, producing an enormous rise in temperature. At a certain stage, this heating causes a generalized explosion which destroys the star and projects the stellar material to huge distances.

In 1971, Ostrikher and Gunn produced a different model: as in Fowler and Hoyle's theory, the collapse of the core creats a neutron star, but this occurs *before* the explosion. The pulsar turns on its axis at great speed (several revolutions per second). Its rotational energy is transferred to the external layers, and this destabilizes the star and produces the explosion. According to Ostrikher and Gunn, the high rotational speed coupled with the extremely large magnetic field of the pulsar (10^{12} G) are enough to explain the 10^{44} J of energy radiated by a supernova.

Many astrophysicists have emphasized the fact that the explosive mechanism cannot be the same for type I supernovae (old stars of medium mass) as it is for type II (young and very massive stars). It has been the general opinion for some years that Ostrikher and Gunn's model would be appropriate for type II but not for type I supernovae. For the latter, C. de Loore has envisaged a completely different process. Consider a close binary formed initially from star A of medium mass (say 5 solar masses) and star B, less massive. When star A has evolved sufficiently to reach the red giant stage, there will be a transfer of mass from A to B due to gravitational effects (see page 190). At the end of a certain time interval, B will become the more massive and all that will remain of A will be a very dense core at high temperature. The inverse effect then begins: because of the higher gravitational pull of A, the transfer takes place from B to A. However, on arriving at A, which is very hot, the material from B will produce a catastrophic explosion of the star.

According to S. Wheeler, this model explains better than others what is observed with type I supernovae, particularly the extraordinary intensity of the X-ray emission. We should add, however, that all the theoretical problems have not been solved and refinement of the theories is still necessary to obtain complete agreement with observation.

THE ROLE OF SUPERNOVAE

The gases arising from a supernova expand over great distances. If it is assumed that a galaxy produces on average one supernova every 50 years, it can be calculated that in a million years every point in our own Galaxy would be swept by gases originating from these gigantic explosions. Since the age of a galaxy is measured in thousands of millions of years, it is true to say that supernovae have a role that we are only now beginning to understand.

The hot gas first of all disturbs interstellar matter over large distances of up to several tens of parsecs. It would seem that 'corridors' of hot gases are formed in this way and are diluted at their boundaries by colder clouds corresponding to normal interstellar matter.

In addition, material from supernovae arriving at high velocities creates a powerful shock wave which locally compresses the interstellar matter. A condensation formed in this way captures surrounding material by accretion, and this is the process that instigates the formation of new stars.

Before and during the explosion, supernovae create heavy elements that are also ejected. These elements, which are missing from less massive stars, are distributed by this process throughout the Universe, which thus becomes gradually enriched in them. In this connection, we should mention a bold but plausible hypothesis that the solar system, which does contain heavy elements, could have been formed by condensation following the explosion of a nearby supernova.

We end by pointing out that supernovae are the main source of cosmic rays, a highly penetrating radiation consisting of particles (protons, electrons, etc.) projected at a speed close to that of light. The role of cosmic radiation in the Universe is still poorly understood but is certainly important.

SUPER EXPLOSIONS

In 1966, B. Westerhund and D. Mathewson reported a gaseous ring in the Large Magellanic Cloud which is the remnant of an explosion. The ring, slightly elliptical, was gigantic: it was more than 1000 parsecs in diameter, in comparison with the gaseous 'loops' of ex-supernovae which barely exceeded 50 parsecs. It has been calculated that this ring has a mass of 100 000 suns!

It is now believed that this is not one star but the whole of an enormous expanse which is exploding: in a region of high stellar density, the conflagration of a supernova would mean the explosion of other stars through heating. This is what some have called a *supersupernova*. Two supernovae remnants have been found in the ring, and in its central region an O–B stellar association can now be seen, that is, a group of young and very hot stars.

This super explosion in the Large Magellanic Cloud is not the only one that is known. Four of them have been identified with varying degrees of certainty in our own Galaxy and we describe each of them in turn.

First of all, there is a hydrogen cloud, discovered by C. Gum in 1952 and now bearing the name of the Gum Nebula. This cloud, which extends over several of the southern constellations, covers 75° by 40°. It forms an elliptic ring, with the enormous dimensions of 700 by 350 parsecs. The centre of the ring is only 450 parsecs from the sun. The mass of the whole structure is estimated at 180 000 solar masses.

In its central part, there is a supernova remnant in the form of a pulsar, PSR 0833-45, whose period is 0.089209 s.*

The Gum Nebula, or rather the whole of the nebulosity, includes a stellar association Puppis O-B I, which contains very hot bright stars such as ς Pup and γ Vel.

A second super explosion seems to have occurred in the region of Orion. A ring has been discovered that extends over 30° in the constellations of Orion and Eridanus and has a maximum diameter of 200 parsecs. Once again, there is an O–B stellar association Orionis OB1 found here together with many hydrogen nebulae, including the well known Orion Nebula, which we shall discuss further in Chapter 5.

A third galactic superexplosion occupies a region centred on γ Cygni. The ring is about 500 parsecs in diameter and includes a certain number of bright and dark nebulae (such as the North America Nebula NGC 7000) as well as a stellar association Cygni OB II which is at the centre. Several X-ray and intense radio sources are also to be found there.

The fourth case is that of a gigantic ring of diameter 1000 parsecs known as the Gould Belt since its existence was suspected by B.A. Gould as long ago as 1874! Its existence has been confirmed by recent work, particularly

---

* This pulsar and that in the Crab Nebula are the only ones to be identified optically with a very faint star (magnitude 24).

by J.R. Lesh (1970), P. O. Lindblad (1973) and O. J. Eggen (1975). The most curious aspect of this is that the centre of the belt is only 200 parsecs from the sun, so that we are *inside* it! A study of the radial velocities of the clouds forming the ring has led two Canadian astronomers, V. A. Hughes and D. Routlege (1972), to date the explosion: it occurred 65 million years ago.

That figure gives rise to a disconcerting theory. About 10 years ago, palaeontologists called attention to a terrestrial catastrophe: at the end of the secondary era the great reptiles that populated the earth, and which are classed under the generic name of dinosaurs,* suddenly disappeared over a period of time that is difficult to estimate but that was ultra-short in geological terms. A large number of marine species also disappeared at the same time. These abrupt disappearances are attributable to a catastrophe which can be dated: it occurred 65 million years ago.

The coincidence between the appearance of a nearby supernova, or of a group of supernovae, and the disappearance of part of terrestrial life is disturbing, even if not accepted as significant by everybody. The supernova was, as we said, only 200 parsecs away and it is now known that the emission of radiation from a supernova is enormous. It is therefore tempting to attribute the disappearance of the large reptiles to the supernova. For the moment, it is impossible to be categorical about this, so let us just say that it is fortunate that human beings did not exist in those far-off times, since they would probably have disappeared at the same time.

We have seen that O–B stellar associations are found in every case of a super-explosion, not as a cause of the phenomenon but as a consequence of it. The shock wave from the explosions stimulates the condensation of hydrogen clouds and hence the formation of stars so that the super-explosion is acting as a catalyst. The O–B stellar associations consist of very young stars and are *subsequent* to the explosion. The process is not finished either: we shall see that stars continue to be formed in regions like Orion and Cygnus, and we now realize what an important role is played in the Universe by the consequences of supernova explosions.

---

* The dinosaurs include many species, often gigantic: tyranosauri, ceratosauri, brontosauri, brachiosauri, stegosauri, etc.

Chapter 3

Novoids
(Nova-like Variables)

Under the general title of novoids (symbol N1), we group together stars with very different characteristics but with the single feature in common that they are eruptive variables. The second supplement to the General Catalogue, which appeared in 1974, created classifications for some of them:

— Gamma Cassiopeiae stars (symbol γ C) consisting of B stars: rapidly rotating subgiants or dwarfs with emission spectra. The variations are often of small amplitude.

— Z Andromedae stars (symbol ZA) or symbiotic stars. These are binaries consisting of a red star similar to a long-period or irregular variable linked to a hot star.

— S Doradus stars (symbol SD). These are supergiants with emission spectra showing irregular and often large variations.

There are also some stars classed as novoids which do not come into one of these three categories: the pseudo-novae or very slow novae, the ex-novae, and several peculiar stars such as η Car.

We now examine each of these groups in turn.

GAMMA CASSIOPEIAE STARS

The star γ Cas was for a long time considered as invariable (visual magnitude 2.20). Its fluctuations were discovered simultaneously by P. Baize and R. Rigollet in August 1936. In April 1937, it passed through a maximum (visual magnitude 1.47) and then declined with small oscillations until May 1938 ($m_V = 2.52$). A short period of growth restored the magnitude to 2.2 and then the decline resumed (m_V 3.0 in 1940). Since then it has oscillated irregularly between 2.2 and 2.7.

This star, of type Be, has a B spectrum with intense emission lines (mainly the Balmer series) which vary continuously in strength, as H. Lockyer showed in 1933. Variations in the line profiles showed a periodicity of 0.70 day. Some very large fluctuations in the emission intensities were observed during the maximum of 1936–37, notably by D. McLauchlin and Tcheng Mao Lin, and at the same time absorption lines appeared.

Another well known Be star is Pleione (BU Tau) in the Pleiades. Its variability was suspected in 1880 by C. Flammarion and confirmed in 1936 by Calder. Its fluctuations, sometimes rapid, are accompanied by significant variations in colour index (Figure 57). The amplitude (0.7 magnitude) is smaller than that of γ Cas.

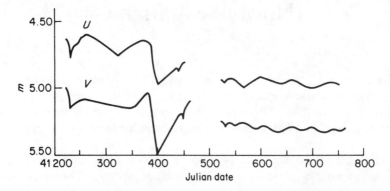

Figure 57.   Light curves of BU Tauri in *U* and *V* from October 1971 to March 1973. (*From A. S. Sharov and V. M. Lyutij*, Peremennye Zvesdy., **18**, *377, 1972. Reproduced by permission of Information Bulletin on Variable Stars*)

Although Be stars are plentiful,* not all of them are known to be variable; Table 32 lists 10 of the brightest which are. The amplitude rarely exceeds 1 magnitude and is often only a few tenths of a magnitude. The spectra

Table 32.   Bright γ Cas variables

Designation	V_{max}	V_{min}	Spectral type
ω CMa	3.68	4.05	B2 IV–Ve
ε Cap	4.44	4.72	B3 IV–Ve
γ Cas	1.47	3.0	B0.5 Vpe
μ Cen	2.92	3.43	B2–3 IVpe
EW Lac	5.01	5.3	B3 III–Ve
χ Oph	4.18	5.0	B2 IVpe
λ Pav	3.4	4.32	B2 Ve
X Per	6.07	6.40	O9.5 III–Vpe
BU Tau	4.77	5.50	B8 IV–Vpe
ς Tau	2.90	3.03	B1.5 IVpe

* 10 to 20% of B stars have an emission spectrum.

range from O8 to B8 with a maximum frequency of occurrence around B2–B3, and it should be noted that the intensity of the emission increases as the star becomes hotter. These stars are subgiants or dwarfs with absolute magnitudes from -2 to -4.

They are young stars that are rotating very rapidly (more than 300 km/s at the equator in the case of γ Cas). In 1942, O. Struve reported that the gravitational force is only just sufficient to keep the material rotating with the star. If this is true, only a slight internal pressure is needed for the equatorial zone of the star to eject material and form an envelope or shell.

Although this ejection could produce some of the photometric variation, it is now believed that this is not the only cause. It was thought that the star could be a binary, but the calculated model did not account well for the observed phenomena. The variation in brightness thus still remains partly unexplained today.

It is interesting that these stars have effectively the same spectra and the same absolute magnitude as the pulsating stars of the β CMa type. Stars that show the two kinds of variation at the same time are now known, so that there is a connection between these two stellar groups. It is, however, a connection that is still not very well understood, and long series of photometric and spectroscopic observations are still needed to establish it conclusively.

Z ANDROMEDAE STARS

Z And is a star that normally varies between magnitudes 10 and 11, with the fluctuations having a semi-periodic form, while at very irregular intervals it can increase to magnitude 8.5. It is the prototype of what is called a symbiotic star, for the emission spectrum of a hot star is superposed on the normal M2 III type spectrum. It is therefore a binary system, but the blue component can only be spectroscopically separated with difficulty because it is very faint compared with the red component. (In the case of R Aqr mentioned in Chapter 3 of Part 2, the B star has a distinctly stronger spectrum.) Figure 58 shows a section of the light curve for Z Andromedae with a faint maximum at Julian date 24 39 930.

About 30 stars of this type are known, of which 8 are listed in Table 33. In some cases, the variations are irregular while others such as AX Per and AG Dra show a certain degree of periodicity. Figure 59 shows the light curves of AG Dra in three colours due to Meinunger, from which it can be seen that the variations are larger in U than in B and V. All these stars have a red giant spectrum (sometimes K but mainly M) linked to a much weaker emission spectrum from a blue star. Bright lines belonging to highly ionized elements (helium, oxygen, iron, etc.) can also be observed in the spectrum.

It is believed that these stars are similar to long-period variables, with the presence of a very hot and variable companion producing perturbations.

122

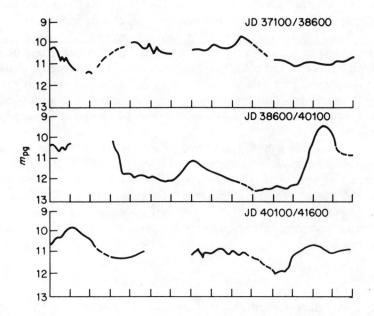

Figure 58. Light curves for Z Andromedae from July 1960 to October 1972. (*From H. Gessner*, Mitt. Veränderlichen Sterne, 7, *No. 3, 1975. Reproduced by permission of Sternwarte Sonneberg, GDR*)

Table 33. Z And type variables

Designation	m_{max}	m_{min}	Period (days)	Spectral type
Z And	8.0	12.4 pg		M2 III + e
CH Cyg	6.7	8.7 V		M7 III + Be
CI Cyg	9.0	13.1 pg	855	M5 III + Bep
V1016 Cyg	11.3	17.5 pg		M3 III ep
AG Dra	9.1	11.2 pg	554	K3 III e +
RW Hya	10.2	11.2 pg	376	M2 III e
AG Peg	6.0	9.4 pg	800	M3 III e + WN6
AX Per	10.8	13.0 pg	685	M5 III ep

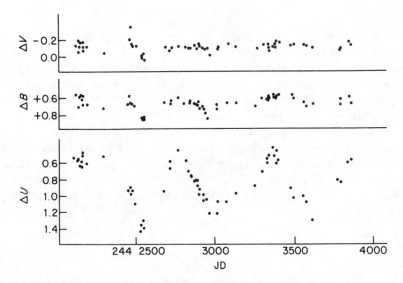

Figure 59.   Light curves in *V*, *B* and *U* for AG Draconis. (*From L. Meinunger*, Inf. Bulletin of Variable Stars, *No. 1611, 1979. Reproduced by permission of the Konkoly Observatory, Budapest*)

In addition, some of them are surrounded by a gaseous nebula which is reminiscent of planetary nebulae. The periodic or pseudo-periodic variations are due to the red component, while the blue star is responsible for the greatest maxima that occur at irregular intervals.

It has been shown recently that two of these symbiotic stars, CH Cyg and EG And, have rapid variations similar to the flickering we have already mentioned, and attributable to the blue component. Figure 60 shows the fluctuations of one of these, CH Cyg.

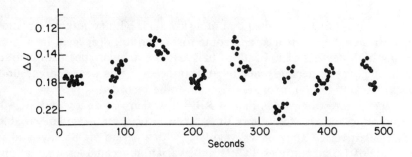

Figure 60.   Rapid variations in CH Cygni. This light curve covers a time of a little over 6 min. (*From J. Mikolajewska and M. Mikolajewska*, Inf. Bulletin of Variable Stars, *No. 1846, 1980. Reproduced by permission of the Konkoly Observatory, Budapest*)

PSEUDONOVAE

A pseudonova or very slow nova is the name given to a star with an eruptive variation of amplitude comparable with that of a nova but sometimes extending over several decades. This is by no means a homogeneous group and we shall therefore confine ourselves to quoting a few individual cases, first of all that of RT Serpentis.

At the beginning of 1909, this star was fainter than magnitude 16, but in July of the same year a magnitude of 13.9 was observed. In 1910, it reached magnitude 11.5 (Figure 61). After that, the increase in brightness became very slow and underwent small fluctuations, but it continued to increase until 1924, when it reached 10.5. The decline then started, with appreciable oscillations. In 1941, RT Ser had a magnitude of 13.5 and it has since fallen to below 15.

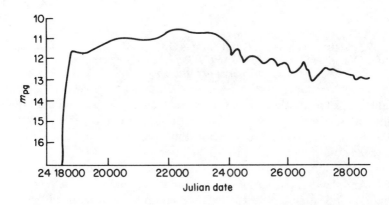

Figure 61.   Light curve for RT Ser between 1908 and 1935

The spectrum first resembled that of a slow nova with an intense emission spectrum including hydrogen and some forbidden lines (oxygen, iron, etc.) showing high degrees of ionization. In 1931, the nebular spectrum characteristic of novae in decline made its appearance and this was observed until the star became too faint for its spectrum to be recorded.

Another interesting case is that of RR Tel, which was discovered in 1908 by Flemming and was for a long time taken to be a semi-regular variable. Until 1930, it varied more or less regularly with a period of 383 days and an amplitude of 1.5 magnitudes. In 1930, the variation became less regular and the amplitude increased. In 1944, the pseudo-periodic variations stopped, and the star which had until then oscillated between 13 and 15.5 magnitudes rose to 12 (Figure 62) and then to 7 in 1948. It remained stable for several years but a decline started in 1958 and still persists. We might add that an examination of photographs at the Harvard photographic library showed that it had already produced a maximum in 1898, but one that was weaker (magnitude 9) and of shorter duration.

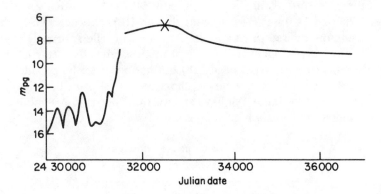

Figure 62.   Light curve for RR Telescopii from 1941 to 1949. Observations taken before the point marked × were made photographically, those after it were visual. (After Ch. Bertaud, *l'Astronomie*, **74**, 121, 1960)

The spectroscopic study of RR Tel was made from 1949 onwards, principally by A. D. Thackeray. At that time, its spectrum was that of an F supergiant with emission lines, similar to that of η Car, which we deal with later. In 1951, the nebular spectrum appeared as with RT Ser, but more recent spectra show a resemblance to the symbiotic stars.

RR Tel, like RT Ser, cannot be considered to be a nova since the phenomenon is too slow to be called an explosion. It is more a continuous but moderate ejection of stellar matter, with the formation of an extensive gaseous envelope responsible for the observed nebular spectrum.

EX-NOVAE

The star V Sge is very different from those we have just looked at: it oscillates around magnitudes 12–13 but sometimes shows periods of high activity during which it increases to magnitude 9.5. In addition, it shows the 'flickering' effect like the novae at their minimum, that is, very rapid variations of small amplitude attributable to flares. Finally, it should be noted that this is an eclipsing binary with a period of 12 h and an amplitude of 0.6 magnitude.

The spectrum resembles that of novae at the end of their 'careers': a nebular spectrum with intense emission lines (hydrogen, ionized helium, highly ionized carbon and nitrogen). The high radial velocities clearly indicate ejection of matter.

V Sge is considered to be an old nova, but it is not the only one. Several others have similar appreciable photometric variations, such as EM Cyg, which varies from 11.9 to 14.4 in *B*. Other stars are quieter and, apart from eclipses, show hardly anything but the flickering (VV Pup and UX Uma). One of the most interesting is undoubtedly MV Lyr: it usually oscillates between 12.1 and 14.0 but it sometimes rises to 10.5. However, G. Romano and L. Rosino discovered in August 1979 that it had fallen abruptly to magnitude 18, only to rise again to 14 several months later. This star is similar to AM Herculis (discussed on page 218) but has a greater amplitude.

Several of these stars are, like novae, short-period eclipsing binaries. Table 34 gives details of 7 of them. All have an emission spectrum resembling that of V Sge, which indicates a high temperature.

Table 34. Some ex-novae known to be eclipsing binaries

Designation	Period			Δm	
	days	h	min		
VV Pup	0.0697	1	40	2.5	*V*
AN UMa	0.0797	1	55	0.6	*B*
TT Ari	0.1326	3	12	0.25	*B*
UX UMa	0.1967	4	44	1.1	*B*
RW Tri	0.2319	5	34	2.0	*B*
EM Cyg	0.2909	6	59	0.25	*B*
V Sge	0.5142	12	21	1.28	*B*

S DORADUS STARS

S Doradus is an odd star. For more than 60 years it remained around magnitude 8.6 with small fluctuations, but occasionally it showed decreases in brightness of about 1 magnitude. The variations were not at all periodic: minima were observed in 1891, 1900, 1930, 1940 and 1955. There was another in 1964, deeper than the rest with the star falling to magnitude 11. The most curious feature is that, this time, it did not rise as usual to its normal maximum but has oscillated ever since then around a magnitude of 10.

The spectrum of S Doradus is that of an A5 supergiant with emission lines similar to that of P Cygni which is described below. S Dor is not a galactic variable but belongs to the Large Magellanic Cloud. It is extremely luminous: taking account of the distance of the LMC, its apparent magnitude corresponds to an absolute magnitude of -10.2, making it as luminous as the most powerful novae at their maxima.

It is a massive star (60 solar masses) but, given the intensity of the radiation from it, it must be evolving rapidly: Wolf estimates that it has lost 1 solar mass in 1500 years. Note also that it has lost 1.5 magnitudes during the century since it was first observed.

Another interesting object classified as an S Doradus star is P Cygni. This was observed in 1600 by Blaeu and was then of the third magnitude. It declined and disappeared to the naked eye between 1620 and 1626, but reappeared in 1655 at a magnitude of 3.5. It oscillated between 4.5 and 5.5 for the whole of the eighteenth century and is now of magnitude 6.

P Cyg has a spectrum (denoted by B1eq) that is peculiar to this star and is known as the P Cyg spectral type. Strong and very broad emission lines are bounded on the violet side by absorption lines due to the expanding gaseous envelope surrounding the star. The radial velocity, also variable, oscillates between 50 and 240 km/s.

P Cyg is very luminous: it is 2100 parsecs away and this gives it a maximum absolute magnitude of −11.9 and a current one of −8.9.

In 1953, E. Hubble and A. Sandage discovered a variable in Messier 31 which oscillated slowly between magnitudes 15.5 and 16.5. Since the distance of this star was known, an absolute magnitude of −9 could be deduced. They also discovered four variables of the same type in Messier 33 (the Triangulum Spiral) fluctuating between magnitudes 16 and 17, giving absolute luminosities of the same order. Eight more were discovered by L. Rosino and A. Bianchini in 1973: 3 in Messier 31 and 5 in Messier 33, one of which has an absolute magnitude of −10.3 at maximum. The mean has been established as −8.8 for M31 and −9.0 for M33. One, AE And in M31, has a spectrum similar to that of η Carinae, to which we now turn our attention.

A CURIOUS OBJECT: ETA CARINAE

This star lies in the southern hemisphere and has been known for a long time: in his celestial atlas of 1603, Bayer estimated it to be of the 4th magnitude. Halley found it to be brighter than that in 1677 (of magnitude 2 we should say today) and several other observers in the eighteenth century attributed the same brightness to it.

From 1830 onwards, η Car was reported as being of the first magnitude, and brighter than the stars of the Southern Cross. Its brightness increased year by year and the maximum was reached in 1843 when it surpassed Canopus and was nearly as bright as Sirius: it was then of magnitude −1 (Figure 63). Its decline was very slow and was punctuated by several secondary maxima in 1856, 1871 and 1889. At the end of the nineteenth century, it stabilized at magnitude 8 but with small oscillations. However, after 1940 it began to increase again and at the moment it is oscillating around magnitude 6.

The star is too far away for an estimation of its distance by trigonometrical parallax. Spectroscopic measurements lead to the assignment of a very high luminosity: during its maximum of 1843, the absolute magnitude was −12.5; between 1677 and 1830 it remained at −9.5; at its minimum it oscillated around −3.5 and it is now about −5.5.

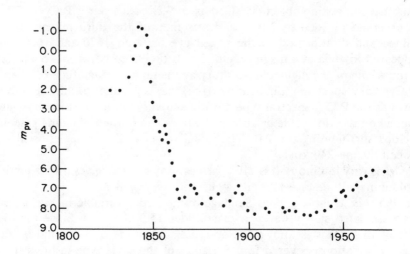

Figure 63.   Light curve of η Car from 1800 to 1970

A calculation shows that in 3 centuries η Car has used up more energy than a type I supernova. A type I supernova may well reach an absolute magnitude of -20, but it rapidly decreases to much smaller values. We also know that a star that has become a supernova has been destroyed, yet η Car seems to be very little affected by the enormous expenditure of energy. This is probably both because the phenomenon is less abrupt and because the mass of η Car is much greater than that of supernovae.

The spectrum shows a large number of emission lines, allowed and forbidden. The speeds of ejection of material are fairly high, of the order of 500 km/s. In addition, η Car is surrounded by a beautiful nebula, NGC 3372, already observed by J. Herschel, and radio wave observations show that there is a ring of expanding matter at its centre.

Infrared measurements, particularly those made by Gehrs, Rey, Becklin and Neugeberger in 1973, show that η Car is an extremely bright infrared object and also that the nebulosity is much more extensive at this wavelength than in the visual or blue: it is nearly 8000 astronomical units across, or 100 times the diameter of the solar system.

Eta Carinae was first seen as a double star in 1914 by Innes and was later seen to be multiple, with the components are moving apart from each other quite rapidly. We have already met such cases among the novae, where it is not so much a matter of doubling as of condensations moving apart from each other.

Many theories have been put forward to explain the origin of such a stellar 'monster' with a total mass that must be greater than 100 solar masses. The most attractive of these to my mind is that put forward in 1976 by P. Maffei, in which Eta Carinae would not be a star but rather a group of massive and very luminous protostars. This group would be similar to, or even bigger than, the Trapezium of Orion, but whereas the stars of the Trapezium have reached a late evolutionary stage, Eta Carinae would be much less advanced. The luminous variations are perhaps due to its enormous nebular envelope which, for more than a century of observation, has undergone appreciable changes.

Eta Carinae may not be unique: G. Tamman and A. Sandage have found a similar one (V12) in the galaxy NGC 2403. Since 1910, this star has fluctuated irregularly between magnitudes 20.5 and 21.5 (Figure 64), but in 1952 and particularly in 1954, it showed maxima which brought it to magnitude 16.5; in other words, the absolute magnitude at that time was −12.3.

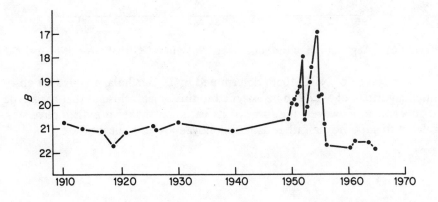

Figure 64.   Light curve of the variable V12 in NGC 2403. Notice the activity and rapid fluctuations in the period from 1949 to 1955

P. Maffei thinks that η Car belongs to the same group as the variables of Messier 31, Messier 33 and NGC 2403: the spectra are similar and the absolute magnitudes are of the same order. However, there is also a suspicion that they could be part of a distinct group of stars such as S Dor and P Cyg.

A SLOW SUPERNOVA

In 1961, P. Wild discovered an object carrying the designation SN 1961v in the galaxy NGC 1058. This supernova, which Zwicky attributed to class V, has not evolved like the others (Figure 65). Its magnitude when discovered was 13.5, but this increased further to 11.5. Its decline was abnormally long (8 magnitudes in 7 years) with oscillations, and the star could be followed until 1970, whereas most of the supernovae take less than a year to lose 8 magnitudes.

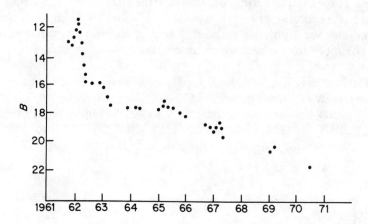

Figure 65.   Light curve of the supernova 1961v in NGC 1038 from 1961 to 1970

The distance of NGC 1058, determined by F. Berthola, enabled an absolute magnitude of −17.2 to be assigned to this object, close to that for a type II supernova. However, the form of its variation makes it a distinctive star, related to supernovae rather as pseudonovae are to typical novae.

Chapter 4

Dwarf Novae

In the three previous chapters on cataclysmic variables, we have described incredibly large and superluminous stars. We now deal with stars that are just as interesting but whose variations in brightness are less spectacular. These are dwarf novae, which are characterized by frequent explosions of limited amplitude. The General Catalogue distinguishes two categories, according to the form of the photometric variations:

— U Geminorum stars: the normal state of these is at their minima. At intervals of time varying widely from one star to another (from ten to several hundred days), there are abrupt maxima with amplitudes ranging from 3 to 6 magnitudes.

— Z Camelopardalis stars: these stars hardly ever stay at their minima; intervals between maxima are usually from 10 to 25 days and the amplitude is from 2 to 4 magnitudes. In some cases, activity stops almost completely and the star undergoes what is called a 'standstill.'

In spite of these observed photometric differences, the two categories seem to form a single physical group.

U Gem, discovered in 1855 by R. Hind, was the first known variable of this type. Normally faint (B magnitude $= 15.0$), it had maxima taking it to magnitude 9. The interval between two maxima can vary a lot (from 60 to more than 200 days), the mean being 103.4 days.

Another well known variable of the same type is SS Cyg, discovered by Wells in 1896. This is the brightest of the dwarf novae (B magnitude 8.2 to 12.5) and one of the best observed of all types of variable. None of its maxima since 1896 (more than 550) have escaped observation. The mean interval between maxima is 50.2 days.

The variability of Z Cam was discovered in 1904 by Christie, but it was only in 1923 that A. Brun reported its peculiar photometric character. It varies from 10.4 to 14.8 magnitudes in B, the mean time between maxima being 23.4 days. However, at irregular intervals it has 'standstills' during which it undergoes very small oscillations around a magnitude of 11.5.

Figure 66 shows the light curves for these three stars.

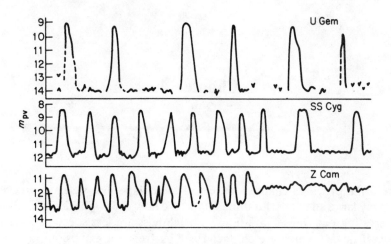

Figure 66. Light curves over 500 days of U Gem (JD 2436800/37300), SS Cyg (JD 2435600/36100) and Z Cam (JD 2441800/42300), from observations of AAVSO and AFOEV

DESCRIPTION OF LIGHT CURVES

Dwarf novae exhibit a great variety of shapes in their light curves. A. Brun and M. Petit in 1954 distinguished seven sub-types (later increased to eight), but this is purely a photometric classification and does not involve any physical characteristics. Nevertheless, a brief description of several of the various kinds of light curve shown in Figures 67 and 68 will be given (note that the horizontal time scale covers 250 days instead of the 500 days in Figure 66).

TZ Per (Figure 67a) has standstills like Z Cam which can be very long; one of more than 700 days has been observed both in RX And and TZ Per, while Z Cam had one from 1976 to 1979 which lasted more than 1000 days. In contrast, CN Ori (Figure 67b) or AH Her, also grouped with the Z Cam stars, show completely different variations. Standstills are rare and of short duration, and some maxima have erratic shapes.

The sub-type X Leonis, which is classed with U Geminorum, is represented here by CZ Ori and UZ Ser (Figures 67c and d). The maxima are almost always asymmetrical, short maxima predominate and the mean cycle is generally between 20 and 30 days.

The sub-type SU UMa, which has a very characteristic shape, is represented by SU Uma and VW Hyi (Figures 68a and b). Most of the maxima are very short with an occasional long and distinctly brighter maximum.

SS Cyg and SS Aur (Figure 68c), together with UU Aql, are characterized by long and short maxima alternating fairly regularly. The amplitude is fairly large (5 magnitudes) and the mean cycle is between 50 and 60 days. The

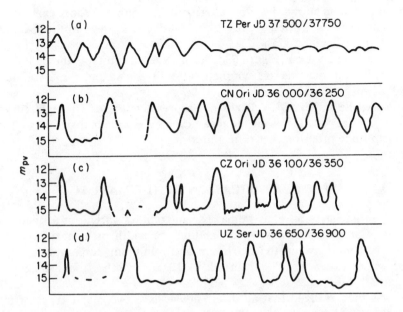

Figure 67.   Light curves of TZ Per, CN Ori, CZ Ori and UZ Ser (from observations of AAVSO, AFOEV BAA and the New Zealand RAS)

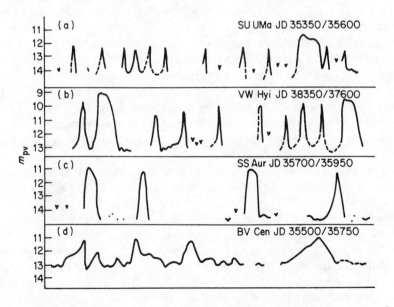

Figure 68.   Light curves of VW Hyi and BV Cen (from the New Zealand RAS), of SU UMa (from observations of AAVSO, AFOEV and BAA) and of SS Aur (from M. Petit)

sub-type U Gem shows the same alternation of long and short maxima, but the mean cycles are often greater than 100 days and the amplitude is often greater than 5 magnitudes.

RU Peg and BV Cen (Figure 68d) form a special sub-type: the light curves are erratic, the maxima are symmetrical or irregular and the amplitude is small (3 magnitudes).

Finally, the sub-type UV Per is not represented here. The light curve is similar to that of U Gem but the mean cycles are greater than 200 days and the amplitude often reaches 6 magnitudes.

THE CLASSIFICATION OF THE MAXIMA

Because these stars have, as we have seen, complex variations at their minima, L. Campbell has envisaged defining the width of a maximum at an arbitrary level between the maximum and minimum; for example, for SS Cyg, which varies visually from 8.2 to 12, the width is defined at visual magnitude 11.

It was then realized that all dwarf novae have two major types of maxima: short maxima, where the decline starts soon after the maximum, which makes them pointed; and long maxima, where the star remains near maximum brightness for several days, giving the maximum a rounded shape. The separation between these two types is always clearcut whatever the star, as is shown by the histograms shown in Figure 69 for the distributions of widths in 6 stars.

Campbell's classification has been extended further. In 1961, Lortet-Zuckermann reported for SS Cyg that, although many maxima are asymmetric (that is, the rise is short compared to the decline) there are also symmetric maxima where the rise is almost as slow as the decline. Two classes of maxima must therefore be considered, long and short, and two sub-classes in each case, normal (or asymmetric) and symmetric. She also revealed the existence of another class, of very short duration and less bright by about one magnitude than the others.

Work on this classification, which at first concerned only SS Cygni, was continued and extended by M. Petit, who has analysed the light curves of about 20 stars.

This has revealed an important point: if the energy E emitted during the maxima is calculated, then for each star the ratio of E for the short maxima to E for the long maxima can be established, and turns out to be generally between 0.30 and 0.40. Similarly, the ratio of E for faint maxima to that for the long maxima is between 0.10 and 0.15. Almost all the dwarf novae show long and short maxima with one important exception, the SU UMa sub-type. The ratios of emitted energies for these stars is close to 0.12 and show that they only have long and faint maxima. They thus form a special sub-type to which we shall return later.

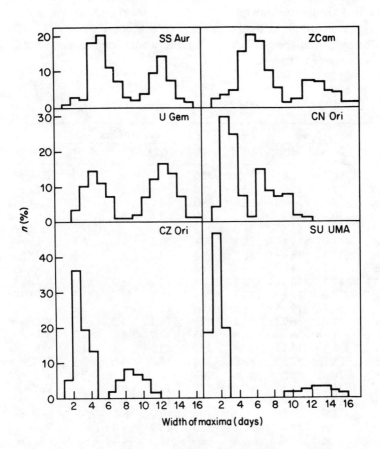

Figure 69. Distribution of widths of maxima for six dwarf novae

The mean cycle (that is, the mean interval between successive maxima) is a convenient figure, but has no physical significance since the means are taken over different types of maxima. Table 35 therefore gives, together with the mean cycle (P_m), the mean interval between long maxima (P_L). Also included are the mean widths of the long, short and faint maxima, and the sub-types according to the Brun–Petit classification.

It can be seen that the SU UMa sub-type, whose long maxima are infrequent, always have a high value of P_L.

Table 35. Dwarf novae

Designation	Δm_B Max.	Δm_B Min.	P_m (days)	P_L (days)	Width of maxima Long	Width of maxima Short	Width of maxima Faint	Sub-type
YZ Cnc	10.2	14.6	11.5	135	13.0		2.5	SU UMa
AB Dra	12.0	15.8	13.8	50	8.6	2.8		Z Cam
WX Hyi	9.6	15.02	14.0	140	10.6		1.5	SU UMa
RX And	10.4	14.0	14.3	60	8.6	4.3		&Z Cam
SU UMa	11.0	14.6	14.6	165	13.2		2.2	SU UMa
TZ Per	11.9	15.6	16.8	75	10.3	5.0	2.6	Z Cam
CN Ori	11.8	15.44	18.1	55	8.5	3.6	2.0	CN Ori
AH Her	10.2	14.7	19.5	62	8.1	3.5		CN Ori
X Leo	11.5	15.5	22.8	67	7.7	3.0		X Leo
AY Lyr	12.6	17.8	23.3	205	14.0		1.8	SU UMa
Z Cam	10.4	14.85	23.4	106	13.7	6.7	2.3	Z Cam
UZ Ser	12.0	16.7	26.4	75	9.1	4.2		X Leo
CZ Ori	11.78	16.2	26.6	98	9.7	3.6	2.2	X Leo
VW Hyi	8.7	14.55	27.8	180	13.3		1.8	SU UMa
SS Cyg	8.2	12.52	50.2	118	14.5	6.8	3.3	SS Cyg
SS Aur	10.5	15.3	54.8	143	12.8	6.1	3.2	SS Cyg
UU Aql	11.0	16.8	60.5	148	11.3	5.0		SS Cyg
RU Peg	10.3	14.0	69.1	110	14.0	5.8	3.0	RU Peg
U Gem	8.9	15.09	103.7	198	12.2	5.1		U Gem
BV Cen	10.5	14.1	105.0	160	13.7	4.0		RU Peg
UV Per	11.6	17.6	367.5	668	12.3	4.3		UV Per

VARIATIONS AT THE MINIMA

At the so-called 'minimum' state, dwarf novae are in fact very unstable. Several different types of variation occur:

Firstly, there are semi-periodic fluctuations clearly visible on the light curves. They have a mean amplitude of 0.05 magnitude and a duration of 5–10 days.

Secondly, there is the flickering that we have already mentioned because it also occurs in other erupting variables: post-novae and pseudo-novae. These are very rapid irregular variations of small amplitude. Figure 70 shows one example.

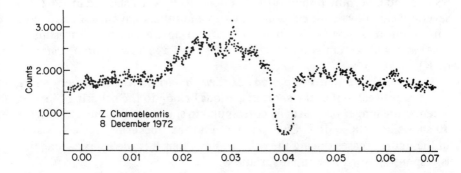

Figure 70. Eclipse of Z Cha observed with a photon counter by B. J. Warner (Sky and Telescope, **47**, *298, 1973. Reproduced by permission of Sky Publishing Corp*)

Finally, B. Warner and his colleagues have revealed the existence of periodic oscillations with small amplitude and durations of several tens of seconds. These are similar to those observed in ZZ Ceti type white dwarfs.

SPECTRA, ABSOLUTE MAGNITUDES AND DOUBLING

The spectra of dwarf novae are generally difficult to observe since the stars are always faint at their minima. They are exceptional in that they have very broad lines. At the minima, it is mainly emission spectra that are observed: the Balmer series of hydrogen, neutral helium and several metallic lines, particularly of ionized calcium. Absorption lines of hydrogen and ionized calcium appear during the rise to the maximum. At the maxima, these are sometimes accompanied by faint emission lines, which increase in strength during the decline while the absorption lines progressively disappear.

The absolute magnitudes are still not very well known. There have been only two determinations by trigonometrical parallax (SS Cyg and U Gem), which have led to absolute magnitudes of about +9, but these were very inaccurate. Various workers, such as Mannino and Rosino, Miczaika and Becker, have used determinations from proper motions and have found still fainter values ($M = +10$ to $+11$), but the proper motions themselves were not very accurate. New determinations on 15 stars by Kraft and Luyten have given a mean value of +7.5. Spectroscopic determinations, carried out mainly by Joy and by Kraft, have given values ranging from +4.8 to +7.5. There is thus considerable uncertainty, but this is to be expected since it is probable that dwarf novae do not all have the same absolute magnitude. RU Peg is known to be fairly bright ($M = 6.5$) whereas WZ Sge is very faint, with an absolute magnitude that appears to be about 10.

It was in 1943 that Joy revealed the spectroscopic doubling of RU Peg and SS Cyg and determined their orbital periods. R. P. Kraft showed in 1962 that several dwarf novae are eclipsing binaries and others have been discovered since then. It is now known that all dwarf novae are binaries, consisting of a hot star with an emission spectrum (dBe) and a yellow or orange star (G or K).

Table 36 shows that the periods of these binaries are always very short, but it is noticeable that the shortest periods belong to the eclipsing binaries, whereas the longer ones usually correspond to spectroscopic binaries. Figure 70 shows an eclipse of Z Cha. The fact that the 'flickering' (clearly visible in the figure) ceases during the eclipse shows that it is the eclipsed star that is responsible for these rapid variations.

Table 36. Dwarf novae known to be spectroscopic binaries (SB) or eclipsing binaries (amplitudes given in the UBV system).

Designation	Period			Δm
	days	h	min	
WZ Sge	0.0567	1	22	0.2 B
V436 Cen	0.0638	1	32	0.1 B
EX Hya	0.0682	1	38	0.9 V
VW Hyi	0.0743	1	46	0.2 B
Z Cha	0.0745	1	47	1.82 V
WX Hyi	0.0749	1	49	0.7 V
TU Men	0.1176	2	49	
BV Cen	0.1580	3	48	0.1 B
U Gem	0.1769	4	15	0.7 B
SS Aur	0.1806	4	20	SB
RX And	0.2117	5	05	SB
SS Cyg	0.2762	6	38	SB
Z Cam	0.2898	6	57	0.1 B
RU Peg	0.3708	8	54	SB

THE MECHANISM BEHIND DWARF NOVAE

The mechanism responsible for dwarf novae has been studied particularly by W. Krzeminski and is similar to what we have encountered in the case of ordinary novae: an orange–yellow star of G or K spectrum is losing some of its material to its companion, a hot star of great density (sub-dwarf or white dwarf).

A hydrogen-rich gaseous ring is formed around the companion (see Figure 71). The arrival of the material at this ring causes a rise in temperature at the point of impact and produces a bright region there, like a luminous spot. From time to time this region explodes and this causes a large variation in luminosity. The phenomenon is more limited than that of an ordinary nova but occurs more frequently. Smak and Warner have pointed out that, in some cases at least, the eclipse is not due to the occultation of the white star by the yellow one, but to its passage in front of the bright spot.

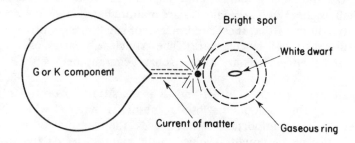

Figure 71.   Processes involved in a dwarf nova

There is one effect not yet accounted for: that of the 'standstills' of the Z Camelopardalis stars. Why do the explosions stop for several months, even several years? No satisfactory explanation has yet been found.

Finally, we report an interesting observation made by F. Cordova and K. Mason on SU Ursa Majoris: the ring of material is, as might be expected, the generator of fairly intense X-ray emission.

FLUCTUATIONS IN THE SU URSAE MAJORIS SUB-TYPE

During the last few years, several observers, notably N. Vogt, B. Warner and J. Paterson, have revealed the existence of a variation which occurs during the maxima of some dwarf novae. It takes the form of a rapid oscillation with an average amplitude of 3 magnitudes, but the oscillation is periodic and its period is practically the same to within 1–3% of the orbital period of the pair.

Vogt and other observers have commented that this oscillation mainly affects the dwarf novae binaries whose orbital period is very short, less than 0.1 day. Now, these stars belong to the SU Ursae Majoris sub-type or to stars resembling them; in other words, to dwarf novae whose long maxima (called 'supermaxima' by some authors) are distinctly brighter than the explosions of short duration. On the other hand, the phenomenon does not occur in the 'classic' U Geminorum stars or in the Z Camelopardalis group.

The theory of the effect is not yet very well established. It is true that the mechanism of the explosion described above is still valid, but it has to be supplemented since one thing seems certain: the different types of maxima (long, short and faint) are probably produced by different processes.

RELATIONSHIP BETWEEN DWARF NOVAE AND OTHER STELLAR GROUPS

Some authors, particularly J. Sahade and R. P. Kraft, have reported a possible relationship between eclipsing binaries of the W Ursae Majoris type and dwarf novae: the latter would be a late stage in the evolution of the binaries.

We shall discuss W Ursae Majoris stars in more detail later, but we should explain straight away that these are binaries of small mass consisting of two main sequence stars (F or G spectra). They are close binaries, called contact binaries, and their periods are always short, generally between 0.3 and 0.6 day.

We shall also see that an effect known as mass transfer occurs in a close binary: one of the components loses some of its material which flows to the other. This has important consequences: the star that becomes more massive in this way gradually evolves to the giant stage. At the end of its life it collapses and is transformed into a white dwarf, with a considerable modification in the orbits.

Thus, the W Ursae Majoris type binary, originally formed from two dwarfs, has become a system consisting of a white dwarf and a yellow sub-luminous star and this is how the dwarf novae binaries are produced. In support of this theory, Kraft has remarked that the W Ursae Majoris and dwarf novae star groups have comparable galactic velocities, that their distributions in the Galaxy are almost identical, and that the two groups both belong to Population II.

However, this theory is not universally accepted. Firstly, it should be pointed out that the measured proper motions which have been used to calculate the galactic motion are imprecise, because they are of the same order of magnitude as the possible error. Secondly, it is not certain that all the W Ursae Majoris binaries are old stars, since they occur in globular clusters, which are not old. Finally, too much credit must not be given, in my view, to the similarity of the galactic distributions because the lack of precision in our knowledge of the absolute magnitudes of the dwarf novae means that their distribution in the Galaxy is not very well known either.

All this leads not so much to the complete rejection of the evolutionary hypothesis of Sahade and Kraft, as to its reception with some caution while awaiting observational confirmation.

In 1934, Kukarkin and Parenago revealed the existence of a relationship between the amplitude A and the mean period P_m which enabled the U Gem stars to be linked to the recurrent novae. Using only 6 dwarf novae and 2 recurrent novae, they obtained the relationship:

$$A \text{ (Vis)} = 1.667 \log P_m + 0.63$$

As observational material became more plentiful, attempts were made to improve this relationship but always in vain. For example, using 63 dwarf novae and 7 recurrent novae, I have obtained (Figure 72) the following alternative relationship:

$$A \text{ (pg)} = 1.50 \log P_m + 2.0$$

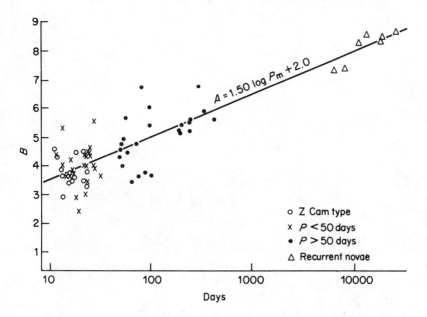

Figure 72. Relationship between amplitude and average interval between explosions for 63 dwarf novae and 7 recurrent novae

Considerable scatter occurs around the mean values whichever relationship is used. There are several reasons for this.

Firstly, the mean period is not a very precise piece of information, as we have seen. The mean interval separating the explosions of the recurrent novae (which have only had 2–5 maxima) are even more inaccurate.

Secondly, the brightness of each one of a pair is not known. In some binaries such as SS Cygni the two components seem to be relatively equal, in others such as U Gem the orange star is almost undetectable, and in a further group such as RU Peg it is dominant. However, precise data on the magnitudes of the separate components are not yet available, so that the *real* amplitude of the explosion is not known. This comment also applies to recurrent novae.

Lastly, it should be pointed out that the spectral types of dwarf novae and recurrent novae differ considerably from each other.

It is therefore possible that we are faced here with a false relationship: an apparent link exists between the two groups but, as Kholopov and Efremov have also indicated, there is no real physical relationship between them. In addition, all efforts to find intermediate stars have so far proved fruitless. Dwarf novae and recurrent novae are not related.

Chapter 5

Variable Stars Associated with Gaseous Nebulae

We now reach an important but complex stellar group: that of variables associated with diffuse nebulae.

Diffuse nebulae consist mainly of hydrogen. They may be bright if the gas cloud is illuminated by one or more stars associated with them; they may be dark if there is no bright star in the neighbourhood or if the opacity of the cloud is too great. In fact, photographs very often show bright and dark regions more or less mixed up together.

We have mentioned that these nebulae contain stars, often a large number of them. They are grouped in 'stellar associations,' which we define below, and some of them are variable. We have therefore to describe, all at the same time, the variables themselves, the stellar associations and the nebulae, and this can demand no little concentration on the part of the reader. The effort is worth making, however, since these gaseous regions are very important: we shall see later that they are the origin of the formation of many stars.

STELLAR ASSOCIATIONS

In 1949, the Soviet astronomer V. A. Ambartsumian revealed the existence of what he called 'stellar associations.' This is a different concept from that of a cluster: a galactic cluster contains stars in greatly varying numbers which are physically connected with each other but which belong to different types. Its dimensions generally lie between 2 and 15 parsecs. A stellar association, on the other hand, contains stars that are *physically similar*, between ten and several hundred in number, and its overall dimensions are much greater.

Ambartsumian recognized two types of stellar association:

— O or OB associations, consisting of very bright O or B stars and generally very large, some extending over 150 or 200 parsecs. We have already encountered this type of association in the chapter on supernovae.

— T associations, so called because the typical star is T Tauri. This stellar group, discovered by A. H. Joy in 1945, has a peculiar type of spectrum: it resembles that of the sun's chromosphere, with emission lines of variable intensity—Balmer lines of hydrogen and metallic lines belonging to neutral or weakly ionized atoms of calcium, iron, titanium, etc. One of the features is the abundance of lithium, 50–400 times greater than that on the sun. We shall return to this feature later. Some T associations are 30–40 parsecs in diameter and may contain several hundred stars, while others are very small (2–3 parsecs).

The stars in these associations are not distributed at random. Some group themselves in the form of a trapezium, the best known of these being that of Orion. Other align themselves along a chain. Finally, Isserstedt and Schmidt-Kaler have shown that in some cases the stars form an elliptic ring 6–7 parsecs in diameter.

CLASSIFICATION OF VARIABLE STARS ASSOCIATED WITH NEBULAE

Given the variety of light curves among these stars, it is difficult to establish a simple photometric classification of them. For all the variables that are associated with diffuse nebulae, the General Catalogue has therefore introduced a classification that uses physical criteria. Three main types are considered.*

In stars

These include irregular variables that are visibly connected to a nebula or located in its immediate neighbourhood. These stars are dwarfs or sub-giants. Several sub-types have been created, and these are briefly described in Table 37.

Is stars

These have the same photometric and spectral characteristics as those of In stars but are not apparently connected with the nebula. Again, several sub-types have been created.

UVn stars

These include flare stars similar to those we shall describe in the next chapter also connected to the hydrogen clouds.

---

* The third edition of the General Catalogue considers a fourth type, Ia, formed from bright stars with an emission spectrum. In fact, however, these are generally classified differently, principally with the γ Cas stars, so that it does not seem very helpful to make a separate class of it.

Table 37. Classification of variables associated with nebulae

Description of star	Type	Sub-type	Comments
Variable connected with the nebulosity	In	Ina	O, B or A spectral types
		Inb	F, G, K or M spectral types
		InT	T Tauri spectral types
		Inas	As above, but with
		Inbs	rapid
		InsT	variations
Variable not connected with the nebulosity	Is	Isa	O, B or A spectral types
		Isb	F, G, K or M spectral types
		IsT	T Tauri spectral types
Flare stars connected with the nebulosity	UVn		K or M spectral types

Table 38 gives details of some of the best known In and Is stars. The T stellar association to which they belong is indicated, together with the nebula that contains them. Several of them (such as RW and UY Aur, or UX and UZ Tau) are binaries with both components being variable.

Table 38. Some In and Is variables

Designation	m_{pg} Max.	Min.	Spectrum	Type	Association and nebula	
RW Aur	9.6	13.6	G5 Ve	IsbT	Aur T1	IC 1579
UY Aur	11.6	14.0	G5 Ve	InbsT	Aur T1	IC 1579
T Cha	10.0	13.8	G5 Ve	Inbs	Cha T1	
R CrA	10.0	13.6	F5 Ve	InbsT	CrA T1	NGC 6729
RU Lup	9.6	13.4	G5 Ve	InbT	Lup T1	B228
R Mon	9.3	14.0	AFpe	InaT	NGC 2261	
T Ori	9.5	12.6	B8–A3 Vep	Inas	Ori T2	NGC 1976
CO Ori	10.3	13.4	Gpe	Inbs	Ori T1	SG 139
SV Sgr	13.4	16.5	dK0e	Inb	Sgr T2	NGC 6524
T Tau	9.6	13.5	G5 Ve	InbT	NGC 1555	
RR Tau	10.2	14.0	A2 II–IIIe	Inas	Tau T4	
RY Tau	8.8	11.1	G2 Ve	InbT	Tau T1	IC 359
UX Tau	10.6	13.3	G2 Ve	InbT	Tau T2	
UZ Tau	11.7	14.9	G5 Ve	InbT	Tau T3	

The variations in brightness of these stars is often irregular and very rapid. An analysis of the fluctuations of some T Tauri stars has shown that they have flares that succeed each other very quickly and become superposed. However, other stars have different variations: in some cases such as T Cha, RU Lup and RY Lup, the fluctuations have a quasi-periodic form. It can be seen that the fact of belonging to types In or Is, that is, to be connected or not to a nebula, introduces no systematic differences in the form of the light curves. This is shown in Figure 73, which reproduces the light curves for two well known stars, RR Tau (Inas type) and RW Aur (IsbT type).

The amplitude of the variations frequently exceeds 2, 3 and even 4 magnitudes but it is generally smaller for stars with earlier types of spectra (Ina or Isa types) than for the others.

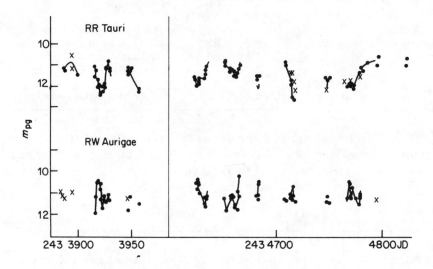

Figure 73.   Light curves of RR Tau (Inas) and RW Aur (IsbT)

THE ORION COMPLEX

This magnificent ensemble, known as the Great Nebula in Orion and very familiar to amateurs, is highly complex.

We have used the expression 'Great Nebula' but there are in fact several nebulae forming a physical grouping. The oldest known of these, which is also the brightest and largest, is Messier 42 (NGC 1976) discovered in 1610 by a student of Galileo, Fabri de Peiresc. It is surrounded by several others, such as Messier 43 (NGC 1984), NGC 1977 and NGC 1999. The whole array covers a vast region of several square degrees corresponding to more then 30 parsecs. The nebulae include an O association called Orion O1 and a T association Orion T2.

In 1656, Huygens noted a group of four quite bright stars in Messier 42 forming a multiple physical system, θ_2 Orionis, called the Trapezium because of its shape (clearly seen in Figure 74). The four stars in this group are as follows:

	V magnitude	Spectrum
θ_2 Orionis A	6.72	B 0.5 V
B	8.1	B 2 V
C	5.13	O7 V
D	6.70	O9.5 V

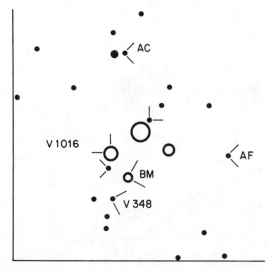

Figure 74. The region of the Trapezium in Orion. The square is of side 90"

A is an eclipsing binary (V1016 Ori) of long period (197.5 days); B is also an eclipsing binary (BM Ori) with a period of 6.470 days, while C is a spectroscopic binary.

There are numerous stars near or within the nebula. G. P. Bond established a catalogue of the region in 1863 that contained 1101 stars. Other catalogues have been published since then, the most important being that of P. P. Parenago (1954), which includes 2982 stars in a region of 9° square. We should add, however, that all are not linked physically to the nebula; some are nearer or more distant stars which lie in front of or behind the nebular formation.

It was also Bond who discovered the first variable star in the region, T Ori. The search for variables became systematic after 1890. The work subsequently carried out is far too extensive to be decribed in its entirety and we mention only two of the most important series of searches. One was undertaken from 1950 onwards at the Tonanzintla Observatory in Mexico by G. Haro and his collaborators; the other was carried out over the same period by L. Rosino and his colleagues at Asiago Observatory in Italy. These two series of observations were continued until the last few years.

The research has certainly been fruitful: we currently know about 780 variables in this region, most of small magnitude (15–18) and often difficult to detect amongst the bright nebulosity. The best known of these variables is T Ori and a small part of its light curve is shown in Figure 75. It can be seen that the variations are completely irregular. The other variables are similar, as can be seen from Figure 76 giving the curves for seven stars observed at Asiago.

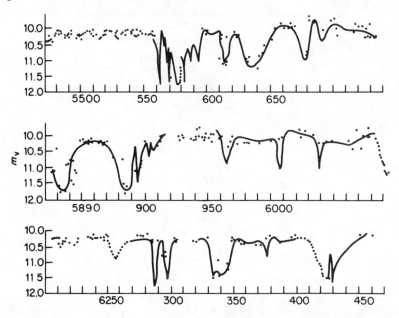

Figure 75.   Light curve of T Orionis (Inas) from 1927 to 1931 (according to G. B. Lacchini, *Asiago Oss. Contributi* No. 70, 1956)

Rosino noted that the light curves differ from each other; some stars remain consistently near the maximum with a minimum from time to time; with others, the converse occurs; for others again, the fluctuations are completely unpredictable, with no preference for increases or decreases.

The Orion region contains a large number of UVn-type flare stars, with about 130 of them being known out of a total of 780 variables. These stars have late spectra from dK5 to dM and are fairly faint, most having minimum photographic magnitudes of 16, 17 or 18 and hence absolute magnitudes from $+8$ to $+10$.

It can be seen from Figure 77 that the amplitude of the flares is generally fairly large, sometimes of 3–4 magnitudes. However, they differ from the flares of typical red dwarfs (dealt with in the next chapter) in that they are less rapid. They often last an hour or two, whereas those of typical red dwarfs of the UV Ceti type sometimes end in a few minutes. For some stars, the flares are even slow.

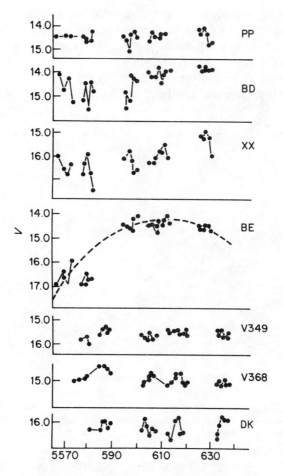

Figure 76. Rapid variations of seven Ins stars in Orion. (From L. Rosino and A. Cian, *Asiago Oss. Contributi*, No. 125, 1962)

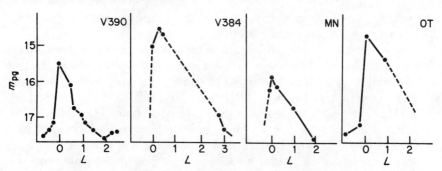

Figure 77. Light curves of four flare stars (UVn) in the Orion region, observed at Asiago Observatory by L. Rosino and his collaborators

It is interesting to study the distribution of variables in the Orion complex. L. Rosino and L. Pigatto have indicated that flare stars are not distributed like In variables. Figure 78 shows that they are distinctly less concentrated in the central part of the aggregate; their distribution is more uniform than that of the In stars. This difference probably reflects a different origin: the stars in the two groups are not 'born' under the same conditions.

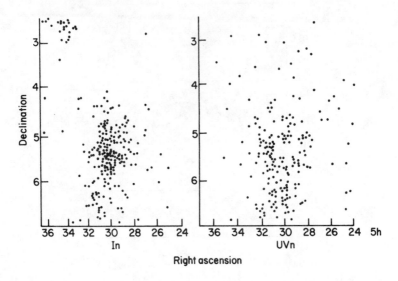

Figure 78.   Distribution of In and UVn variables in the Orion region. (From L. Rosino and L. Pigatto, *Asiago Oss. Contributi*, No. 231, 1969)

SOME OTHER NEBULAR REGIONS

It is not possible for us to describe or even quote all the formations similar to that just dealt with. Table 39 gives details of 12 of them, chosen from among the largest and best observed. The number of known variables in Table 39 is given only as an indication of the relative concentrations, since discoveries are still being made every year.

We have said that T-associations are of various sizes, some of them having diameters of 20, 30, 40 parsecs or even more, and sometimes containing more than 100 variables. On the other hand, there are small associations such as Cephei T1 and others still smaller. In some cases, nebulae called cometoids have been detected with the shape of a comet's tail: NGC 2261, containing only 3 variables (of which one is R Mon) is like this, and so is NGC 6729, a nebula with 7 variables, one of which is R CrA.

Among the important stellar associations, we can quote Monoceros T1, which includes the nebula NGC 2264: it is vast (40 parsecs across) and contains more than 200 variables. Then there is Sagittarius T2, studied first

its increase in brightness: it was then of type M4. Then, at the beginning of 1979, just after its maximum, it had become type A4. One year later, it was found to be F5, with hydrogen and sodium emission lines and many metallic absorption lines. The radial velocities are fairly small (30–40 km/s).

We thus have a very special class of star: in a few months or a year, it gains 6 or 7 magnitudes, which corresponds to a ratio of 250 between the maximum and minimum brightnesses. It then loses between 0.5 and 1 magnitude. At the same time, its spectrum passes from a late K or M type to type A at maximum and type F slightly later, implying a large increase in effective temperature. Finally, the small measured radial velocities show that there is no question of an explosion.

What may appear surprising are the size and great speed of the change in the spectra. We shall shortly meet the explanation which has been given for this.

HERBIG–HARO OBJECTS

We must say a little about some special objects, the first of which were discovered in 1954 by G. Herbig and G. Haro in the region of the Great Nebula in Orion. These workers noticed considerable changes on photographs taken between 1947 and 1954: certain small nebulosities visible in 1947 had become fragmented 7 years later and had assumed a quasi-stellar aspect.

Since that time, a large number of Herbig–Haro objects have been discovered and studied in various regions. In all cases, the changes have been rapid. Observations made at various wavelengths, but particularly in the infrared, makes us think that what is being observed is a gaseous envelope surrounding a protostar that is still very young and for the moment invisible.

STAR FACTORIES

Evaluation of the age of clusters and associations has been made possible through an evolutionary theory established by the Japanese astrophysicist Hayashi in 1961. We shall not go into this complex theory in detail, since it is outside the scope of a book of this size. It is enough to say that the positions of cluster and association stars in the H–R diagram are a function of their age. These stars have therefore only to be located in the H–R diagram, and this is possible if good measurements of their magnitudes are available in several colours so as to give precise colour indices. The age of the formations can then be estimated.

It has been shown in this way that O and T associations are, as we have said, young or very young objects. The oldest of them have not been in existence for more than 10 million years[*] and some associations are even

---

[*] This is very small in comparison with main sequence stars, which are generally much older (5000 million years for the Sun).

younger, for instance Cygni T1 (1 million years) or Monoceros T1, associated with the nebula NGC 2264, to which M. F. Walker has assigned an age of 400 000 years. Some stars inside associations are even younger: 10 000 years or less.

The presence of an element such as lithium in the spectrum is very important. The atomic nucleus of lithium is very fragile and is destroyed by nuclear reactions taking place in stellar interiors. The presence of large quantities of lithium in the atmosphere shows that a star has not yet reached the stage of nuclear reactions. It is therefore not a star in the astrophysical sense of the word, but a *protostar*; in other words, an object still at the contracting stage. This contraction produces heat which, according to Hayashi, is removed by convection in the interior of the protostar, whose radiated energy is therefore of gravitational and not nuclear origin.

It is inferred from this that T Tauri objects are not yet stars but protostars. Observations, made particularly by M. Cohen and L. V. Kuhl (1979), lead us to think that they have fairly small masses, of about 0.2–0.3 solar mass, but that their diameter is large (from 1 to 5 solar diameters) since they are not completely contracted. A protostar is often surrounded by a ring of nebular material and also by a cloud of silicate 'granules' which are stable at those temperatures (1000 K). Observations in the infrared region carried out over several years have revealed the complex nature of the phenomena occurring around these stars.

In many regions of the sky, small dark globules can be observed that are easy to detect since they form dark spots against the background of bright nebulae. They were discóvered in 1947 by B. J. Bok and E. Reilly in Messier 8, and since then a large number have been observed in nebular formations. We now begin to understand the structure of these objects, whose dimensions lie generally between 0.3 and 1 parsec. At the centre, there is a protostar (or even several); on the outside, a sort of 'cocoon' masks the protostar, absorbs its radiation and re-emits it in the form of infrared radiation. The globule thus constitutes a primitive pre-stellar stage.

Infrared observations have revealed the existence of objects much more compact than the globules. These have been known for some 20 years, but observations made using satellites, and particularly the IRAS, have enabled a large number of them to be detected, especially in the region of Orion, in the cloud surrounding Rho Ophiuchus and in a large star cloud, Chameleontis I, which contains no less than 70 infrared sources. These objects are interpreted as protostars already at a more advanced stage than those inside the globules. As for T Tauri itself, this represents the last stage of a protostar before the initiation of nuclear reactions.

Where can the FU Orionis type objects be placed? Some workers, e.g. G. H. Herbig, originally thought that the considerable increase in the luminosity of these stars was the result of illumination from nuclear reactions in the core, in other words, of the passage from the protostar to the mature star. However, the presence of considerable quantities of lithium in their atmospheres contradicted this hypothesis since we have seen that lithium is destroyed by nuclear reactions. Another more plausible theory has therefore been put forward: at an advanced stage of the contraction, the temperature of the protostar is high enough to dissociate the molecules in the cocoon that surrounds it. This cocoon collapses in a few weeks or months, and when it has disappeared it is the much more luminous protostar that we observe. The increase in brightness is thus interpreted as being due to the breaking up of the cocoon. After that, the protostar shows only small variations.

Nebular regions are the seat of extremely complex phenomena, one of which is of particular interest to us: the interaction between the gas and the stars, which is not yet very well understood. It is nevertheless responsible at least in part for the variations in brightness of the stars. In some cases, using radio waves, 'jets' of material have been detected that are responsible for appreciable radio-wave emission.

There is also the fact that, as we have seen, these nebulosities provide us with valuable information about the formation of stars and the evolution of very young stars. Lastly, there is another interesting feature that we shall say only a little about since it lies outside our subject: radioastronomical observations made with millimetre waves have shown that clouds also 'man-ufacture' *organic* molecules. More than 50 of these have now been identified, and among some of the very complex ones are those necessary to life.

Chapter 6

Variable Red Dwarfs

Red stars with low intrinsic luminosity (dwarfs with K or M spectra) frequently show variations in brightness. The variations may be of two types: rapid and irregular flares, capable of reaching several magnitudes but of short duration, the typical star of this group being UV Ceti; or periodic variations of small amplitude due to the presence of a small bright region or spot on the star, the typical star here being BY Draconis.

Some of the flare stars are associated with clusters and their characteristics are slightly different from those of typical UV Ceti stars. They form a third group which we shall also discuss.

DISCOVERY OF THE UV CETI STARS

These are among the faintest known stars and only those very near to us can be detected. Almost all those now known are in fact at least 25 parsecs from the sun. The two first stars of this type known were discovered by A. Van Maanen at the Mount Wilson Observatory: the first in 1940, a faint companion of the star BD + 44° 2051 (which received the designation WX UMa); the second in 1943, Ross 882 or YZ CMi. It is curious that these discoveries remained almost unnoticed for several years. It needed W. J. Luyten (1948) to observe a flare in the B star of a small binary L 726–8, since called UV Cet, for the phenomenon of flares to become interesting.

The binary L 726–8 is formed from two red dwarfs whose spectra dM5.5e and dM6e show strong emission. Both stars are variable, the B star appearing to be the most active, and any observations made generally refer to the pair without it being possible to say with certainty which has produced a flare.

Discoveries of UV Ceti stars became more numerous in the 1950s and at least 120 are now known within a radius of 25 parsecs around the Sun.

156

Table 40 gives details of those which are less than 5 parsecs away. It will be noticed that V645 Centauri is none other than Proxima, the star with the greatest known parallax. Three stars have been added that exceed the

Table 40. UV Ceti stars situated less than 5 parsecs from the Sun

Designation		V_{min}	Spectra	r (parsecs)	M_V
V645 Cen		11.06	dM5e	1.30	15.49
CN Leo		13.50	dM6.5e	2.37	16.66
UV Cet	A+B	12.01	dM5.5e	2.58	14.95
V1216 Sgr		10.62	dM5.5e	2.90	13.31
FI Vir		11.10	M4.5V	3.35	13.47
EZ Aqr		12.18	dM5.5e	3.45	14.49
GX And		8.08	dM2.5e	3.45	10.39
GQ And		11.06	dM4.5e	3.45	13.37
YZ Cet		12.04	dM5.5e	3.83	14.12
DO Cep		11.32	dM4.5e	3.95	13.34
V577 Mon	A+B	11.10	dM4.5e	4.06	13.05
FL Vir	A+B	12.53	dM5.5e	4.27	14.38
TZ Ari		12.27	dM5e	4.46	14.02
V1581 Cyg	A+B	12.91	dM5.5e	4.74	14.53
DY Eri		11.17	dM4.5e	4.83	12.75
AD Leo		9.43	M4Ve	4.85	11.00
EV Lac		10.20	dM4.5e	5.00	11.71
WX UMa		14.45	dM5.5e	5.38	15.80
V1298 Aql		17.48	dM5e	5.85	18.64
YZ CMi		11.20	dM4.5e	5.99	12.31

5 parsec limit by a small amount: WX UMa, because it was the first flare dwarf discovered; V1298 Aql, because it is the faintest known; and YZ CMi, because it is one of the best observed stars of this type.

We have already mentioned that these stars are very faint: the brightest of them has an absolute magnitude of +8, but several have absolute magnitudes of +15 to +18: in other words, they radiate 10 000, 100 000 or even 200 000 times less energy than the Sun. Many of these variables belong to a binary system and in several cases both components are variable. If the pair are some distance from each other so that they can be individually observed, they receive two designations (GQ and GX, for example); if, on the other hand, the components are closely bound and difficult to observe separately, they receive only a single designation (such as UV Ceti).

Close binaries are of great importance: a determination of their orbits enables the masses of the components to be obtained. The UV Ceti stars, very faint red dwarfs, are the least massive stars known. Table 41, giving details of several binaries of this type, shows that some of them have masses that are less than a tenth of a solar mass.

158

Table 41. Some UV Ceti stars. The column headed *P* gives the period of revolution of the pair. When the pair is a close binary, the spectral type is the mean of the two stars. Note particularly the pair ADS 3093C, in which one component is a white dwarf and the other a red dwarf

Designation of binary	Designation of variables	Spectral type A	B	M_V A	B	P (yr)	Mass (Sun=1) A	B
ADS 246	GX and GQ And	dM2.5e	dM4.5e	10.39	13.37	3030	0.26	0.13
ADS 433	A = V 547 Cas	dM2.5e	dM4	10.51	12.32	320	0.40	0.13
Kruger 60	B = DO Cep	dM4	dM4.5e	11.57	13.34	44.5	0.27	0.17
L 726-8	A + B = UV Cet	dM5.5e	dM5e	15.46	15.96	26.52	0.109	0.105
ADS 3093BC	C = DY Eri	DA	dM4.5e	11.11	12.73	252.2	0.45	0.20
Bz 200	A + B = AT Mic	dM4.5e	dM4.5e	11.36	11.45	675	0.16	0.16
Ross 614	A + B = V577 Mon	dM4.5e		13.10	16.45	16.60	0.114	0.062
Wolf 630	A + B = V1054 Oph	dM3.5e		10.77	10.81	1.714	0.42	0.42
Bz 237	A+B = EQ Peg	dM4e	dM5e	11.33	13.35	177.9	0.32	0.22
Wolf 424	A + B = FL Vir	dM5.5e		15.01	15.2	16.2	0.068	0.085

OBSERVATION OF FLARES

From 1950 onwards, several observers undertook a systematic surveillance of UV Ceti and a few other stars of the same type. Photographic obervation proved to be very difficult because of the short duration of flares and the great speed of the rising part of the variation. A visual watch was therefore kept, which was tedious since the unpredictability of the variations mean that they had to be observed continuously. Nevertheless, the watch bore fruit and several dozen flares were seen on UV Ceti, one of which was of more than 5 magnitudes.

It is only since 1966 that various observatories have undertaken systematic observation using photoelectric photometers, either with one colour or several colours simultaneously (generally *U*, *B* and *V*). In addition, programmes have been combined with radio-wave observations or with some in the far-ultraviolet or X-ray regions using satellites. It has thus been possible to obtain an idea of the process responsible for the variations. We should add that the number of flares observed in UV Ceti exceeds 1000 and has reached several hundred for other variables, notably AD Leo, YZ CMi and EV Lac.

Flares occurring in the same star can take different forms, as is shown by Figure 80 for UV Ceti. The most common form is that of No. 12, while

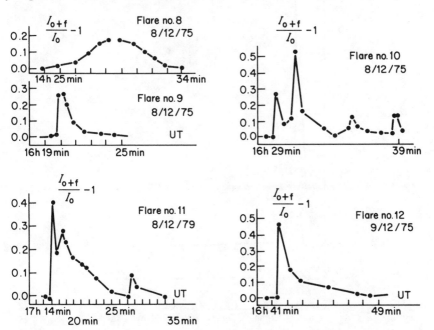

Figure 80. Five flares in UV Ceti observed in *B*. The amplitudes are expressed in terms of luminous intensity with the minimum being taken as the base value. (*From B. B. Sanwaal*, Inf. Bulletin of Variable Stars, *No. 1210. 1976. Reproduced by permission of the Konkoly Observatory, Budapest*)

some are preceded or followed by a secondary flare (Nos. 10 and 11). Slow flares, as in No. 8, also exist. A similar variety is found in other stars.

With rapid flares, the rise to the maximum takes place in a few seconds. Thus, a flare of 3.5 magnitudes has been observed in UV Ceti in which the rise time was only 6–7 s. The decline is significantly slower, but in most cases does not last longer than 10 min.

The amplitude varies greatly from one flare to another in the same star, and it also depends a great deal on the colour. Figure 81 for YZ CMi shows the differences in amplitude according to whether the observation is made in U, B or V. This is in complete agreement with what we know of spectral changes. We also understand why a flare observed in V is smaller than in B, which is in turn smaller than that in U: small variations that are difficult to detect in V are easier to observe in the shorter wavelengths.

In some cases, the existence of what may be called anti-flares has been noted. The variation is as abrupt as that in flares but is a decrease in brightness instead of an increase. An example is shown in Figure 82.

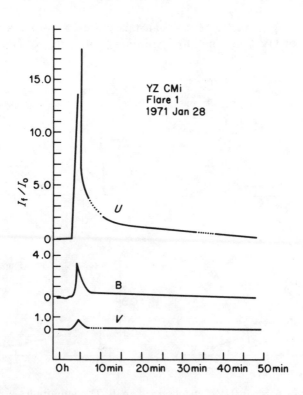

Figure 81. Luminous intensities in U, B and V of a flare in YZ Canis Minoris. (*From S. Cristaldi and M. Rodono*, Inf. Bulletin of Variable Stars, *No. 554, 1971. Reproduced by permission of the Konkoly Observatory, Budapest*)

The distribution of amplitudes for a given wavelength can be studied and the example chosen is that of UV Cet shown in Figure 83. In 1975, the observatories at Pretoria (South Africa), Uttar Pradesh (India) and Thessalonika (Greece) carried out a surveillance lasting 132.5 h. Out of 77 flares observed (in B), 3 had an amplitude greater than 3 magnitudes, 31 were between 1 and 3 magnitudes and 43 were less than 1 magnitude.

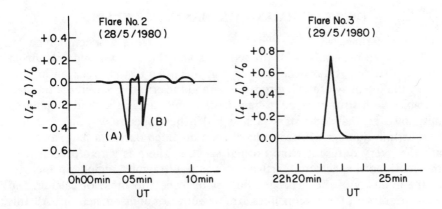

Figure 82. Flare and antiflare in V774 Herculis. (From F. M. Mahmoud and M. A. Soliman, *Inf. Bulletin of Variable Stars*, No. 1866, 1980). While flare No. 3 is normal, No. 2 is an antiflare, shown by its sudden decrease in brightness. (*Reproduced by permission of the Konkoly Observatory, Budapest*)

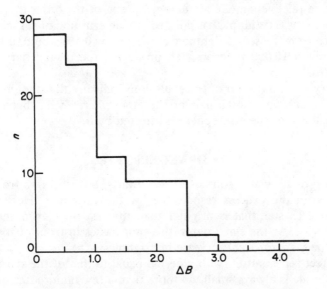

Figure 83. Amplitude in B of 77 flares in UV Cet

For a given star, the frequency of flares varies with time. Sometimes they succeed each other in rapid succession and at others they are separated by long periods. It also varies from one star to another; various workers (W. Kunkel, 1970, M. Petit, 1970; R. E. Gershberg and M. Shakhovskaya, 1971) have shown that there is a correlation between the frequency of flares and the luminosity of the star. The weakest stars are the most active, while in the relatively bright stars such as CR Dra flares are very infrequent.

THE SPECTRA AND MECHANISM OF FLARES

During the times when no flares occur, UV Ceti has a spectral type M, characterized generally by hydrogen lines and those of ionized calcium in emission (dMe). It has been noticed that there is a distinct relationship between the intensity of the hydrogen lines and the eruptive activity of the star; variables with strong emission lines have many flares. It should be pointed out, however, that some stars such as FI Vir have spectra without emission.

A certain number of spectra have been obtained during a flare, and they are then very different from normal spectra. There is a reinforcement of the continuous spectrum, particularly in the ultraviolet region, an increase in the intensity of the hydrogen lines with respect to those of calcium and an appearance of emission lines from neutral or ionized helium. All this indicates a considerable rise in temperature. The colour temperature also changes, which explains why the amplitude appears greater in B than in V and greater in U than in B.

A flare is an explosion produced in the atmosphere of a star, analogous to the eruptions that can be observed daily in the chromosphere of the sun. It has a relatively moderate energy output, of the order of 10^{25}–10^{26} J. Nevertheless, as F. Whipple has pointed out, an eruption of sufficient energy to produce an increase in brightness of the sun by 0.1% would, if transposed to a red dwarf 10 000 times weaker, produce a spectacular flare of several magnitudes.

The very short duration of the explosions indicates that the phenomenon is localized in a very small part of the atmosphere; it has been estimated that only 1–3% of the atmosphere is affected by it.

BY DRACONIS STARS

This group of stars also consists of red dwarfs, but it differs from the UV Ceti group in the process responsible for the variations. Here there is a region on each star that is brighter than the rest of it, in other words, a luminous spot. As the star rotates, this spot passes in front of the observer and produces an apparent increase in brightness.

The effect repeats itself with a period equal to that of the star's rotation. The amplitude is always small, no more than a few hundredths of a magnitude, and the variation can only be detected by photometric observation.

About 25 of these stars are known. Table 42 gives details of seven of them, of which five are spectroscopic binaries.* Table 42 therefore gives both the spectroscopic period (time of revolution of the pair) ($P_{spect.}$) and the photometric period (time for one rotation) ($P_{phot.}$), but it should be noted that the two are often equal and generally between 2 and 10 days.

Table 42. BY Draconis variables

Designation	V_{max}	$P_{phot.}$ (days)	$P_{spect.}$ (days)	Spectra	M_V
EQ Vir	9.34	3.96	Single	K5 Ve	7.3
BY Dra	8.23	3.8	5.976	dM0e	7.49
YY Gem	9.21	0.814	0.814	M0.5 Ve	8.40
CC Eri	8.86	1.561	1.561	K7 Ve	8.7
FF And	10.38	2.170	2.170	M0 Ve	8.7
AU Mic	8.75	4.865	Single	dM2.5e	9.18
CM Dra	12.90	1.267	1.267	dM4e	12.03

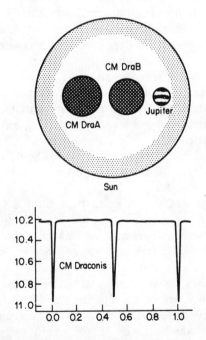

Figure 84.   The binary CM Draconis, according to C. Lacy (1977). Above: comparative sizes of the components, the Sun and Jupiter. Below: eclipses of CM Dra. Note the variation at the maxima, due to BY Dra type phenomena

---

* It is likely that the interaction between the two close components plays a significant role in the formation of the luminous spots.

Two of the spectroscopic binaries, YY Gem and CM Dra, show eclipses. These are the two least massive eclipsing binaries known; the mass of the YY Gem pair is 1.30 solar masses and that of CM Dra is 0.45 solar mass. Figure 84 illustrates the latter two.

BY Draconis stars are often more luminous than UV Ceti stars, most of them having absolute magnitudes between $+7$ and $+8$. The spectral type often lies between dK5 and dM0 but, as with UV Ceti, emission lines from hydrogen and ionized calcium are often observed.

It is important to point out that most BY Draconis stars, especially those in Table 42, also have flares so that they behave at the same time like UV Ceti stars. On the other hand, some stars classed as UV Ceti stars also have a periodic variation of the BY Draconis type. This is the case with YZ CMi, whose photometric period is 2.78 days. In the end, therefore, it is difficult to separate the two types.

FLARE STARS IN GALACTIC CLUSTERS

We saw in the previous chapter that many flare stars are observed in regions that are rich in hydrogen clouds, particularly the Great Nebula in Orion. An appreciable number have also been detected in young galactic clusters. The most typical case is that of the cluster in the Pleiades (Messier 45) familiar to amateurs. A systematic search for flare stars in this region began in 1964 at the Tonanzintla (Mexico) and Asiago (Italy) Observatories. It was continued in other observatories and is still operating.

The search has proved extremely fruitful as regards the Pleiades region with the detection of several hundred very faint variables in a few square degrees. Many other regions have also been studied, especially that of the Praesepe cluster in Cancer (NGC 2632) where about 50 have been discovered. The cluster of Coma Berenices (Melb. 111) should also be mentioned, where about 35 are known. The older clusters are less rich: only eight flare stars are known in the Hyades.

All these stars are generally very faint; thus, the Pleiades variables have on average an absolute magnitude of around $+11$. Te spectral types range from dK5 for the brightest to between dM4 and dM5 for the faintest.

Given their low intrinsic luminosity, these stars are detectable only in galactic clusters relatively near to us; at 100 parsecs, a star of absolute visual magnitude $+11$ has an apparent magnitude of 16 visually and 17.5 photographically; at 400 parsecs, the figures would be 19 and 20.5.

The flares are slightly different from those of the typical UV Ceti stars, and resemble those of the variables in the nebular regions. The amplitude frequently exceeds 3–4 magnitudes in B and 6–7 magnitudes in U. The rise is rapid, but the total duration of the phenomenon is often more than 1 h, as can be seen in Figure 85, which relates to four stars in the Pleiades. Lastly, the frequency of the flares is also lower than for the UV Ceti stars.

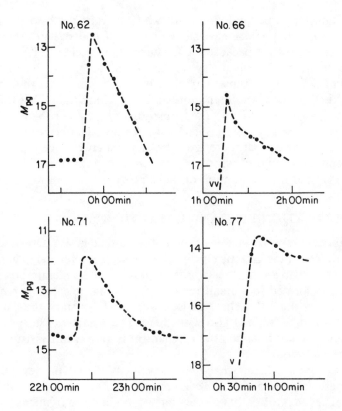

Figure 85.   Flares in four stars in the Pleiades (L. Pigatto and L. Rosino, *Asiago Oss. Contributi*, No. 246, 1971)

SOME COMMENTS ON THE PHENOMENON OF FLARES

A lot of work has been carried out on the subject of flares and we can only summarize a little of it very briefly. We mention particularly that of W. Kunkel, V. Oskanian, R. Gershberg and their collaborators, which has revealed a number of physical relationships. The most important of these are the following.

Firstly, there is a correlation between the absolute magnitude and the frequency of the flares: the least luminous stars are the most active. This relationship was briefly stated in 1970 by M. Petit and was subsequently established more precisely by W. Kunkel.

Kunkel has also shown that there is another relationship between the absolute magnitude and the duration of flares: stars of very small luminosity have flares that last for shorter times than those of more luminous stars. There is, on the other hand, no clear correlation between the amplitude and the duration of flares; weak flares lasting a long time are sometimes observed, as are large amplitude flares of short duration.

The relationships just quoted explain why the variables associated with open clusters or with nebulae have slower and less frequent flares than those of the UV Ceti stars. It is because the former are, on average, brighter stars.

Being a member of a binary system seems to be an important factor; clearly, some single stars show flares but it is noticeable that the spectroscopic or visual binaries are often more active. It seems likely that the presence of a nearby star encourages the formation of flares. For the pair V1054 Oph, with one of the shortest periods known for a visual binary (1.71 years), Kunkel (1975) showed that there is a relationship between the frequency of the flares and the orbital phase: the explosions are more frequent when the components are at their minimum distance apart.

POPULATION TYPE AND GALACTIC DISTRIBUTION

All the stars encountered in this chapter, UV Ceti stars, BY Draconis stars and the variables in the galactic clusters, have one characteristic in common: the existence of flares—in other words, explosions of moderate power and short duration located in a small region of the atmosphere, but recurring frequently. They share this characteristic with the UVn stars, the flare stars associated with nebulae, and also with the T Tauri stars whose spectra are of an earlier type but in which the mechanism is known to consist of many superposed explosions.

All these variables form what it is convenient to call the 'evolving sequence'; these are young stars (or even protostars like the T Tauri) in rapid evolution. They very often have emission spectra and are frequently associated with great expanses of gas.* They closely follow the spiral arms of the Galaxy and hence belong to a young Population I.

It is noticeable that the oldest galactic clusters do not contain flare stars, that those of medium age have a few of them and that only the young clusters include a large number. There is thus a correlation between the age of the clusters and the number of flare stars they contain.

To end this chapter, we point out that some UV Ceti variables are older than others. This is so for those whose spectra do not show emission lines, such as FI Virginis. They belong to an intermediate population, but they do not seem to be very numerous.

---

* It should be pointed out that many of the youngest galactic clusters contain nebulosities. This is notably the case with the Pleiades.

Part 4

ECLIPSING BINARIES

Double stars play an important part in stellar astronomy. We recall briefly (since this is a little outside the scope of this book) that they fall into three large classes.

First, there are the visual binaries, that is, pairs whose components are optically separable. A large number of these are known (60 000 or more), but most of them are not well characterized since observation over long periods of time is needed to calculate their orbits. There are only 700 whose orbits have been determined and of these fewer than 200 have elements that could be considered really certain.

The second class is that of the spectroscopic binaries: if the two components of the pair are very close, they cannot be resolved optically. They are revealed instead by the superposition of their spectra; depending on the respective positions of the components relative to an observer, the spectral lines of the pair either separate or merge together.

The first spectroscopic binary to be identified was Mizar (ς UMa), discovered in 1890 by W. Pickering. Several thousand are now known but not all have been completely studied. The last General Catalogue, established in 1978 by H. Batten and his colleagues, brought together 968 with calculated orbital elements. A supplementary catalogue followed in 1984, edited by A. Pedoussaut, and this contained about 200 additional stars.

The third class, that of eclipsing binaries, is a special case of the second. If the line of sight of the observer is reasonably near the orbital plane of the spectroscopic binary, an eclipse occurs when one component passes in front of the other. Figure 86 explains the effect: if the secondary star is some distance from the principal one, we receive the sum of the emitted radiation. When the principal star is eclipsed, there is an appreciable decrease in

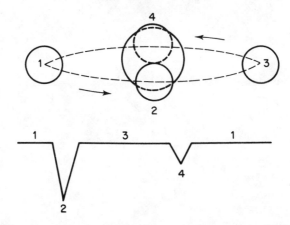

Figure 86.   Eclipsing binaries. At positions 1 and 3, the secondary star is a great distance from the principal star and the brightnesses add, giving a maximum. At position 2, the principal is partially eclipsed, giving a principal minimum. At position 4, the secondary is eclipsed by the principal component, giving a secondary minimum

luminosity and this is called the principal minimum. When the secondary is eclipsed, there is again a decrease in brightness, but a smaller one, and this is the secondary minimum.

Eclipsing binaries, which are also sometimes called 'photometric binaries' or 'geometric variables,' are numerous. The General Catalogue lists about 5000 of them. We should add that the recent catalogue edited by H. K. Brancewicz and T. Z. Dworak (1980) gives the orbital elements of 1048 eclipsing binaries.

The first chapter in this section of the book deals with the mechanism of the eclipses, with the various types of eclipsing binaries and with their characteristics. The second chapter is devoted to an examination of special effects and of those eclipsing binaries where one component is an intrinsic variable.

Chapter 1

Eclipsing Binaries:
Description of Types

A great variety of eclipsing binaries exists, not all with the same type of light curve. The General Catalogue considers three large classes, designated respectively by the symbols EA, EB and EW, while if the class is undetermined the notation used is simply E.

Class EA, whose typical star is Algol, is the most widespread and consists of binaries in which the two stars revolving around each other are at a large distance apart. Eclipses occur with a deep principal minimum, a shallow secondary minimum and relative stability in brightness outside the eclipses. This is the first case illustrated in Figure 87.

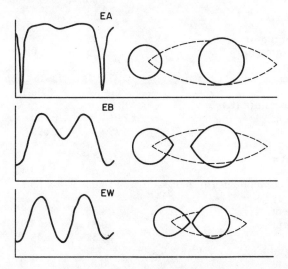

Figure 87.   On the left: light curves of eclipsing binaries of classes EA, EB and EW. On the right, diagram of the corresponding pairs of stars

Class EB, whose typical star is β Lyrae, is very different: the two stars, generally giants or supergiants of low density and of unequal size, are not very far from each other. Because of their mutual attraction, they take on an ovoid shape, and their light curve is therefore rounded, very often with an appreciable secondary minimum. This is the second case in Figure 87.

Class EW, whose typical star is W UMa, is similar to the previous group except that here the stars are two dwarfs, almost equal in size and brightness. They are very close to each other, even in contact. The two components are highly deformed by their attraction and the light curve takes the form of the third case in Figure 87; the secondary minimum has almost the same amplitude as the principal minimum.

The variations in Algol were discovered in 1669 by Montanari. It was Goodricke in 1784 who suspected that the variations in brightness might be due to eclipses, a hypotheses that was daring for its time. This was confirmed in 1890 by W. Pickering.

It was Goodricke again who discovered the variability of β Lyrae, the prototype of group EB, but its nature as an eclipsing binary was not recognized until the twentieth century.

Table 43.   The three typical stars of the eclipsing binaries (The data concerning temperatures, radii, separation of the components, masses and absolute magnitudes are due to H. K. Brancewicz and T. Z. Dworak, 1980)

Characteristic	Algol	β Lyr	W UMa
Periods (days)	2.8674	12.9081	0.3336
Maximum magnitude (V)	2.12	3.34	7.88
Amplitude:			
Principal minimum	1.28	1.00	0.75
Secondary minimum	0.06	0.50	0.66
Spectral type:			
A component	B7.5 V	B8 II–IIIep	F8 Vp
B component	G8 IV	B6.5	G2 Vp
Effective temperature (K):			
A component	12 010	10 260	5 920
B component	4 290	10 690	6 140
Radius (Sun = 1):			
A component	2.74	11.10	1.09
B component	3.60	4.43	0.80
Separation of components (in solar radii)	14.03	55.51	2.58
Mass (Sun = 1):			
A component	3.70	9.85	1.30
B component	0.81	3.94	0.76
total	4.51	13.79	2.06
Absolute magnitude:			
A component	−0.6	−3.0	+4.5
B component	+3.5	−1.2	+5.0

The history of W UMa, of class EW, is much shorter. It was recognized as a spectroscopic binary in 1920 by W. Adams and A. H. Joy, and observed to be an eclipsing binary in the same year by H. Shapley and H. van der Bilt.

The photometric characteristics of the three types of binary and the physical properties of their components are gathered in Table 43. It can be seen that β Lyrae is a pair of large stars, whereas the components of W UMa are similar to the Sun as regards their spectra, radii, masses and luminosity. Algol, on the other hand, has two very unequal components, the principal one providing 95% of the total emitted energy. This accounts for the faintness of the secondary minimum.

Having dealt with the three types of eclipsing binaries, we now describe each of the groups in more detail.

BINARIES OF GROUP EA

The EA binaries, also known as Algol variables, occur in large numbers; the General Catalogue lists about 3500 of them. They belong to spectral types from O to M and include not only giants and supergiants but also dwarfs. However, the most frequently occurring spectra are those of type A, followed by B and F.

Table 44 gives details of 21 of these stars chosen from among the brightest of them. Columns Δ_1 and Δ_2 give the amplitudes of the principal and secondary minima, while columns D_1 and D_2 indicate the duration of these minima expressed as a fraction of the period. The masses of the pairs are also given in solar masses. It can be seen that the duration of the secondary minimum is generally very short, often less than a hundredth, or even a thousandth, of the period.

There are several main types of light curve. First, there is the Algol type we have already described, in which the secondary minimum is very small because of the great difference in luminosity of the components. Many variables are like this, for example AR Cas, U Cep, δ Lib and X Tri, for which Figure 88 shows the light curves in two colours obtained in 1952 by F. Lenouvel. The second component is sometimes so faint that the secondary minimum is practically undetectable (AW Peg, for example).

The second case is that of pairs formed by stars that are roughly equal in size and brightness. The secondary minimum is then almost as large as the principal minimum, as the example of Y Cyg in Figure 89 shows.

The third case is that of stars formed from one bright star of average size and a fainter star of great size. The principal minimum is then wide since the occultation of the principal star by the secondary lasts a long time. The secondary minimum is not very deep. This is the case with very long period binaries which we discuss later.

Table 44. Some bright type EA eclipsing binaries (upper section, systems consisting of dwarfs or subgiants; lower section, systems including a giant or supergiant). The mass is that of the whole binary system in solar masses

Designation	P(days)	V_{max}	Δ_1	Δ_2	D_1	D_2	Spectra	Mass
R CMa	1.136	5.70	0.64	0.08	0.16	0.000	F1 V + K1 IV	1.38
RZ Cas	1.193	6.18	1.04	0.08	0.17	0.000	A2 V + G1 IV	2.50
ς Phe	1.670	3.94	0.50	0.30	0.12	0.00	B6.5 V + A0 V	9.09
δ Lib	2.327	4.92	0.98	0.09	0.28	–	A0 V + G1 IV	3.60
U Cep	2.493	6.80	2.30	0.08	0.15	0.039	B7 V + G8 IV	6.70
WW Aur	2.525	5.72	0.73	0.58	0.10	0.000	A7 V + A7 V	3.55
β Per	2.867	2.12	1.28	0.03	0.14	0.000	B7.5 V + G8 IV	4.51
λ Tau	3.958	3.41	0.48	0.09	0.15	0.000	B3 V + A4 IV	7.49
β Aur	3.960	1.90	0.10	0.09	0.06	0.00	A2 IV + A2 IV	4.60
AR Aur	4.134	6.15	0.67	0.55	0.07	0.000	B9 V + B9 V	4.85
YZ Cas	4.467	5.66	0.40	0.07	0.15	0.043	A2 IV + F4 V	4.10
BM Ori	6.470	6.38	0.57	0.04	0.12	0.04	B2 V + A5 V	13.01
RR Lyn	9.945	5.64	0.39	0.3	0.04	–	A7 V + F3 V	2.75
α Cr B	17.370	2.24	0.11	0.03	0.034	0.00	A0 V + G3 V	3.75
U Sge	3.381	6.58	3.60	0.13	0.14	0.019	G2 III–IV + B9Ve	9.58
AR Cas	6.066	4.84	0.12	0.04	0.06	0.00	B8 III + B9	12.03
V380 Cyg	12.426	5.68	0.12	0.08	0.11	–	B1 III + B3 V	26.06
μ Sgr	180.45	4.34	0.13	–	0.11	–	B8 ep Ia	24.35
ζ Aur	972.16	3.75	0.13	0.12	0.041	0.039	B6 V + K1 Ib	13.93
V1488 Cyg	1147.4	3.98	0.3	0.03	0.02	–	K3 Ib II + B	30.80
V695 Cyg	3784.3	3.77	0.11	0.06	0.017	0.015	B6 V + K4 Ib–II	15.14

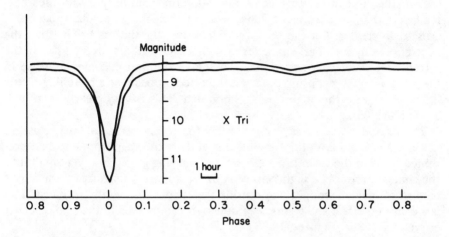

Figure 88.   Light curves in *B* and *V* of X Tri (F. Lenouvel)

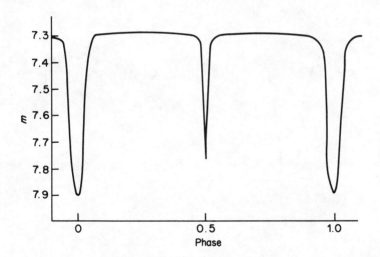

Figure 89.   Light curve of Y Cyg

176

We digress to describe two effects observed in a large number of binaries. The first is called 'limb darkening' and is clearly observable on the Sun. At the centre of the star, where the atmosphere of the star is traversed through the smallest thickness, we receive the radiation from deeper and thus hotter layers than that from the edges where the thickness of the atmosphere crossed is greater. On the Sun, the edges are slightly less bright than the central region. Such darkening varies with wavelength; it is very small in the infrared, still not very pronounced in the visible, but becomes more so in the ultraviolet. In eclipsing binaries, the effect appears as a rounding of the light curve, with the angles being smoothed. This is very clearly seen in X Tri (Figure 88).

The second effect is less obvious: when the components are fairly close, as in classes EB and EW, they assume a slightly ovoid shape and are then presented to us in the form of an ellipsoid of varying area. However small it is, the change produces a slight variation in the apparent magnitude; the curve, instead of being a straight line outside the eclipses, is slightly rounded. This is a common effect and a good example of it can be seen in the photoelectric curve of RT Per (Figure 90).

Algol variables have extremely varied periods, ranging from less than 0.1 day to 10 000 days or more. We have constructed a histogram showing the

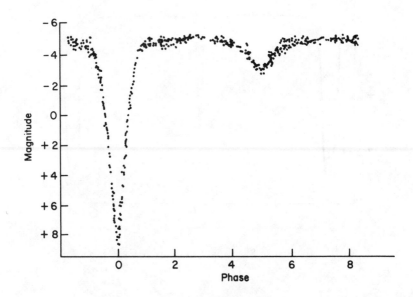

Figure 90.   Light curve of RT Persei (class EA). (*From S. Mancuso and L. Milano, Inf. Bulletin of Variable Stars, No. 1102, 1976. Reproduced by permission of the Konkoly Observatory, Budapest*)

distribution of the known periods of 2802 stars (Figure 91), from which it can be seen that the very short and very long periods are rare whereas there are a great number between 0.5 and 10 days, the modal value being between 2.5 and 3 days.

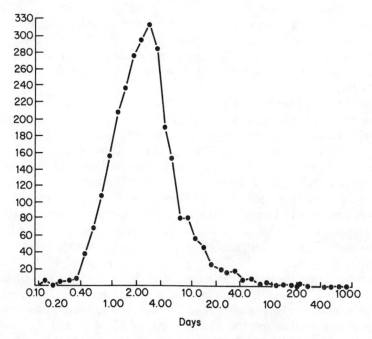

Figure 91. Distribution of the periods of 2802 EA eclipsing binaries between values of 0.1 and 1000 days

We first take a brief look at the very short period variables. The first one discovered was UX UMa, having a period of 0.1966 day or 4 h 43 min, and consisting of two faint white stars; it is now known to be an old nova. Some of the dwarf binaries are in fact eruptive variables and several of them have been dealt with in the chapters on novae (Table 28), novoids (Table 34) and dwarf novae (Table 36).

However, there are two short-period binaries of more importance than most which must be mentioned. They are the central stars at the nucleus of planetary nebulae, in other words, stars at the end of their evolution. Their importance stems from the fact that there are very few binaries among the nuclei of planetary nebulae.

The first is UU Sge, which is surrounded by the planetary nebula Abell 63 and was observed by M. G. Bond. It has a short period of 0.4650 day and is formed from an extremely hot O subdwarf and a K dwarf. Its photographic magnitude varies between 14.7 and 19.0; such a large amplitude shows that the small but luminous subdwarf is totally eclipsed by the secondary component.

The other case is that of the binary V477 Lyr surrounded by the nebula Abell 46. The spectrum is not known but the period is almost the same as that of UU Sge, i.e. 0.4710 day. The amplitude, on the other hand, is much smaller at 1.4 magnitude.

We now pass on to the very long period binaries and spend more time on these. A dozen are known with periods of more than 1000 days and nearly all of these are presented in Table 45. Some consist of a B star as the principal component, linked to a K or M supergiant; ς Aur, AZ Cas and V695 Cyg are examples. In others, such as V777 Sgr or V1488 Cyg, the supergiant is the principal star.

We deal briefly with those binaries which have the longest known periods. First, there is VV Cep with a period of 7430 days. It had minima in 1957 and 1977, and the next will begin in April 1997 and end in August 1998, lasting a total of 491 days. This is one of the most massive binaries known, of nearly 100 solar masses, and the principal component has a diameter 1621 times greater than that of the Sun, that is, 2270 million km! If it replaced the Sun in the solar system, it would extend as far as a point between the orbits of Saturn and Uranus. We shall see in the next chapter that it also undergoes intrinsic variations. (Another star resembling this one is KQ Pup with a small amplitude but, according to A. Cowley, with a period of 9752 days.)

The other binary is ε Aurigae whose period is even longer at 9892 days, or more than 27 years. Its latest minima occurred in 1929, 1956 and 1983, and the next will take place in 2010. In fact, the last minimum took 700 days, with the decline and the rise each taking 170 days and the minimum proper lasting 360 days; it began in July 1982 and only finished in June 1984. The enormous size of the orbit and the duration of the minimum had led to the assumption that the secondary component has the extraordinary dimensions of the order of 3000 solar diameters (or 4000 million km). However, infrared observations have not detected this star, whereas it should be easily detectable. Two hypotheses are now being considered, one in which the companion is a hot dwarf (perhaps surrounded by an accretion disc) and the other in which it is a cold disc of particles revolving around a star emitting very little radiation. A study of the observations made at the time of the last eclipse should enable the veil to be lifted from this mysterious star.

Stars may exist with even longer periods, but they have not been sufficiently observed up to now for us to be certain about them. First there is BM Eri, an M spectrum star, which had a minimum in 1944–45 lasting for more than 1 year but none since. It is assumed that its period is equal to, if not greater than, 50 years. Then there is V644 Cen; the light curve of this star could be equally interpreted as that of a RCrB-type variable, but if it is indeed an eclipsing binary, D. O'Connell suggests that its period must be close to 200 years since the minimum has lasted more than 10 years.

Table 45. EA eclipsing binaries with very long periods. The magnitudes of AZ Cas and V381 Sco are photographic

Designation	P(days)	V_{max}	ΔV	D_1	Spectra	Masses (solar units)			Radii (solar units)		M_V
						M_1	M_2	Total	R_1	R_2	
V777 Sgr	936.07	9.4	0.2	0.06	K5 Ib + A	13.42	5.54	18.96	109.11	4.23	−4.4
ς Aur	972.16	3.75	0.15	0.041	B6 V + K1 Ib	8.60	5.33	13.93	3.49	168.84	−4.5
V1488 Cyg	1147.4	3.98	0.3	0.02	K3 Ib–II + B	22.64	8.16	30.80	350.77	3.91	−5.2
EE Cep	2049.5	10.73	1.8	0.13	B5 II–IIIe +	–	–	–	–	–	−4.0
AZ Cas	3404	11.0	1.0	0.03	B0e + M0 Ib	18.96	15.91	34.87	10.89	280.25	−6.5
V695 Cyg	3784.3	3.77	0.11	0.017	B6 V + M4 Ib–II	9.23	6.18	15.41	6.62	168.23	−4.8
V381 Sco	6545	12.3	3.7	0.10	A5 Ia + M5 Ia	25.18	19.17	44.35	155.21	105.12	−7.0
VV Cep	7430	6.65	0.81	0.06	M2ep Ia + B8 Ve	63.81	35.09	98.90	1621.36	22.24	−6.5
KQ Pup	9572	4.88	0.29	–	M2ep Ia + B	–	–	–	–	–	−5.6
ε Aur	9892.0	2.94	0.89	0.08	F8 Iae +	–	–	–	–	–	−8.5

BINARIES OF GROUP EB

We have seen that the rounded nature of the light curve of these stars is due to the elliptical shape of the components. Since they are frequently of the similar spectral types and thus of comparable brightness, the secondary minimum is often pronounced. The stars are generally giants or supergiants, with not very late spectra, mainly O or B but also A and F. The masses are often large since we are dealing with stars of high luminosity.

Table 46 gives details of ten β Lyrae stars, chosen from among the brightest. It can be seen that they all belong to spectral types O and B and that they all have high luminosity. Figure 92 reproduces the light curve in two colours of one of the most luminous, UW CMa, with absolute magnitude -7.4.

At present, some 600 EB variables are known. The histogram in Figure 93 shows that their periods range from 0.4 to nearly 200 days and that they are fairly well concentrated around 4 days, with a marked modal value of between 0.8 and 1 day.

Long periods are rare; only three are known which exceed 100 days and these are for large-sized supergiants: ν Sgr (137.94 days), BM Cas (197.28 days) and W Cru (198.53 days).

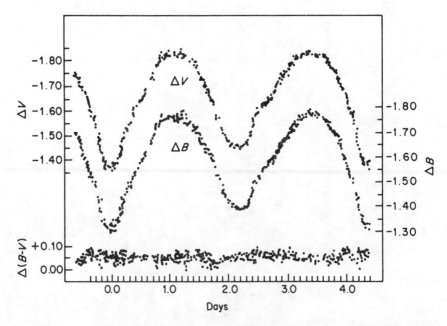

Figure 92.   Light curve of UW Canis Majoris (class EB). The amplitudes are measured with respect to a comparison star. (*From H. Ogata and C. Hukusaku,* Inf. Bulletin of Variable Stars, *No. 1235, 1977. Reproduced by permission of the Konkoly Observatory, Budapest*)

Table 46.   Bright EB eclipsing binaries

Designation	P (days)	V_{max}	Δ_1	Δ_2	Spectra	Masses (Sun = 1)			M_V
						M_1	M_2	Total	
μ_1 Sco	1.440	3.02	0.28	0.11	B1.5 Vp + B6 V	13.56	8.94	22.50	−4.3
VV Ori	1.485	5.44	0.37	0.17	B1 V + B4 V	15.26	5.34	20.60	−4.9
σ Aql	1.950	5.20	0.10	–	B3 V + B3 V	5.70	4.90	10.60	−3.3
u Her	2.051	4.83	0.68	0.26	B3 III + B5 III	7.21	2.88	10.09	−3.1
SZ Cam	2.698	6.84	0.30	0.25	B0 II–III + B0	24.01	6.96	30.97	−4.5
AO Cas	3.523	5.90	0.14	–	O9.5 III+O9.5 III	31.73	26.02	57.75	−6.9
UW CMa	4.393	4.48	0.45	0.35	O7f + O9 III	32.62	24.47	57.09	−7.4
η Ori	7.989	3.14	0.21	–	B0.5 V +	12.76	12.13	24.89	−5.6
β Lyr	12.908	3.34	1.00	0.50	B8 II–IIIpe + B6	9.85	3.94	13.79	−3.4
ν Sgr	137.94	4.51	0.10	0.05	B8 pI + F2 pI	21.26	18.10	39.36	−4.1

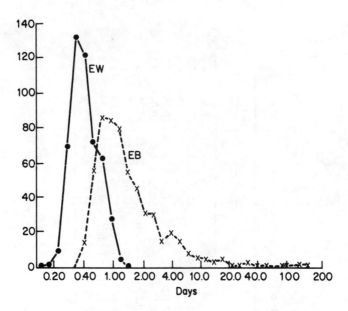

Figure 93. Distribution of the periods of 503 class EW and 578 class EB eclipsing
binaries

BINARIES OF GROUP EW

This stellar group consists mainly of dwarfs, belonging mostly to spectral
types dF and dG. The two components are often comparable in size and
brightness, with the secondary minimum almost as deep as the principal
minimum.

Table 47 gives the properties of ten of the best known of these stars. Apart
from W UMa, the prototype star, whose B and V light curves are shown in
Figure 94, we include i Boo because it is the closest eclipsing binary to the
sun (13 parsecs).

About 500 variables of this type are known. Their periods are always
short and Figure 93 shows that their distribution is very different from
that of the EB stars. There is a very marked modal value between 0.3
and 0.4 day, and periods longer than 0.8 day are rare. It should be noted
that there is a correlation between the spectral type and the period, with a
transition between K spectra for the shortest periods to A spectra for the
longest.

The EB stars may be young, but the EW stars are generally considered
to be old stars belonging to Population II. However, the presence of four
of them (EP, EQ, ER and ES Cep) in a galactic cluster of medium age,
NGC 188, shows that this may not be exactly true. Some EW stars certainly
belong to Population II, but there are others that undoubtedly should be
allocated to the intermediate population.

Table 47. EW eclipsing binaries

Designation	P (days)	V_{max}	Δ_1	Δ_2	Spectra	Masses (Sun = 1)			M_V
						M_1	M_2	Total	
i Boo	0.268	5.85	0.59	–	G2 V + G2 V	0.96	0.48	1.44	+5.2
VW Cep	0.278	7.8	0.41	0.34	K0 V + G5 V	0.82	0.38	1.10	+6.0
YY Eri	0.321	8.8	0.70	–	G5 V + G5 V	0.99	0.61	1.60	+4.4
W UMa	0.384	7.88	0.75	0.66	F8 V + G8 V	1.30	0.76	2.06	+3.9
U Peg	0.375	9.23	0.57	0.51	G2 V + G2 V	1.36	1.09	2.45	+3.0
ER Ori	0.423	9.33	0.68	0.66	G3 V + G3 V	0.62	0.38	1.00	+4.1
OO Aql	0.507	9.12	1.00	0.9	G5 V + G5	0.91	0.59	1.50	+3.9
ε CrA	0.591	4.79	0.26	0.21	F2 IV + F3 V	3.27	0.36	3.63	+3.2
RR Cen	0.606	7.27	0.41	0.4	F5 + F9	1.65	0.35	2.00	+2.3
S Ant	0.648	6.45	0.52	0.56	A9 V + F4 V	0.68	0.33	1.01	+1.5

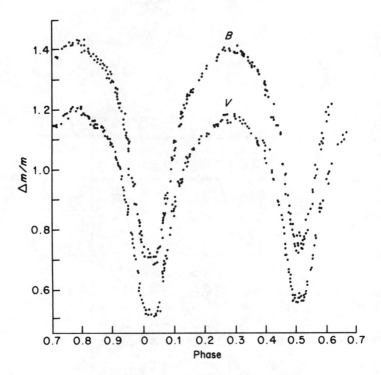

Figure 94.   Light curves for W Ursae Majoris. (*From S. Evren, S. Tunca and O. Tumer*, Inf. Bulletin of Variable Stars, *No. 1577, 1979. Reproduced by permission of the Konkoly Observatory, Budapest*)

ELLIPSOIDAL VARIABLES

The General Catalogue has created a separate class for a dozen stars called ellipsoidal variables. These are binaries whose components have an ovoid shape but whose plane of revolution makes a large angle with the direction of our line of sight. There are therefore no eclipses, but merely a variation in the apparent stellar area. This produces a slight variation in brightness (a few hundredths of a magnitude) detectable in bright stars. Among these, we may cite η_5 Ori, whose period is 3.70 days and whose amplitude is 0.05 magnitude, ψ_2 Ori ($P = 2.526$ days), IW Per ($P = 0.917$ day), IX Per ($P = 1.326$ days), o Per ($P = 4.419$ days) and b Per ($P = 1.527$ days). These are all stars with B or A spectra concentrated in the galactic arms (Orion and Perseus), and they occur in OB associations or young galactic clusters. They therefore form a typical Population I.

DETERMINATION OF ORBITAL ELEMENTS

Like all double stars, the eclipsing binaries provide valuable information for astrophysicists, particularly about their stellar masses that are so important in the study of stellar evolution. Because so much work has been devoted to the determination of the orbital elements of these systems, it is not possible to describe it all here. We shall only mention that of the theoretician Z. Kopal, who has specialized in this field of study.

Although it is also impossible to give details of the method of determining the orbit, which would require formidable mathematical techniques, we shall nevertheless try to give a brief indication of the method of proceeding.

The first step is to construct a *corrected* light curve; in other words, one from which secondary distortions have been removed (limb darkening, ellipticity of components, etc.). The depth of the two types of minimum first enables the *relative* luminosity of the two components to be established. If we denote the radius of the orbit (or the semi-major axis) by a, the relative values R_1 and R_2 of the radii can then be calculated. These are R_1/a and R_2/a. A photometric study also enables the inclination i of the system to be calculated.

To pass to absolute values, an additional parameter is required; the distance to the Sun. Since the parallax is unknown (except in a few cases), the spectroscopic parallax is used together with the curve of the radial velocities of the two components. With these data, the absolute value $a \sin i$ is calculated, and as i (and hence $\sin i$) are known, the value of a can be determined. Since the relative radii R_1/a and R_2/a are known, the absolute values of R_1 and R_2 can be obtained. Knowing the orbital radius and the distance, the masses M_1 and M_2 can be determined, together with the densities since the stellar radii are known.

Significant progress has been made in our knowledge of the absolute values for orbital elements. In 1940, S. Gaposchkin published a catalogue giving the elements for 244 eclipsing binaries. The last catalogue, that of Brancewicz and Dworak, contained 1048 of them.

THE MASSES OF THE BINARIES

We have mentioned that it is important to know stellar masses, and this is especially true for young massive stars since visual binaries are never suitable for the determination of the orbits of very bright stars; they are always a long way apart and have very long periods.

Table 48 gives the properties of twelve of the most massive of the eclipsing binaries known. There are a number of O and W spectra stars among them;[*] it can be seen that the O stars have a mass of between 20 and 40 suns and the Wolf–Rayet stars between 15 and 20 suns. Table 48 also includes several long period supergiants such as VV Cep or V381 Sco. Table 48 includes several binaries of small mass (less than 2 suns) apart

---

[*] As an indication, we mention that V382 Cyg, a short period supergiant, has one of the greatest orbital velocities known, 710 km/s.

Table 48. Masses and radii of eclipsing binaries. The upper part gives details for 12 very massive stars (more than 40 solar masses); the lower part for 8 of those with the smallest known masses outside the EW type

Designation	Type	P (days)	V_{max}	ΔV	Spectra type	Masses (solar units)			Radii (solar units)		$M_1 + M_2$
						M_1	M_2	Total	R_1	R_2	
GP Cep	E	6.688	8.96	0.11	B0 III + WN6	23.95	16.12	40.07	9.73	4.08	−5.9
V381 Sco	EA	6545	12.3	3.7	A5 Ia + M5 Ia	25.18	19.17	44.36	155.21	105.12	−5.5
V729 Cyg	EB	6.598	9.05	0.32	O7 f + O8	26.70	23.00	49.70	8.17	7.07	−7.4
EM Car	EA	3.414	8.73	0.30	B0 I + O5 V	26.80	23.25	50.05	8.79	7.03	−7.9
V444 Cyg	EA	4.212	8.3	0.30	O8 III + WN5.5	34.53	19.03	53.56	10.32	2.28	−7.5
UW CMa	EA	4.393	4.45	0.45	O7 f + O9 III	32.62	24.47	57.09	13.16	10.34	−7.4
AO Cas	EB	3.523	5.90	0.14	O9.5 III + O9.5 III	31.73	26.02	57.75	6.48	5.89	−6.9
V453 Sco	EB	12.004	6.36	0.37	B0.5 Iae + B1 Iab	31.04	27.94	58.98	9.44	11.16	−5.7
V382 Cyg	EB	1.885	9.01	0.78	O7.5 I + O9 I	37.16	32.79	69.95	8.07	7.90	−6.7
UU Cas	EB	8.519	10.4	0.4	B0.5 + B4	48.37	36.27	84.64	26.83	23.10	−7.0
VV Cep	EA	7430	6.65	0.81	M2ep Ia + B8e	63.81	35.09	98.90	1621.36	22.24	−7.5
RY Sct	EA	11.525	7.34	1.50	B0ep I +	52.49	47.24	99.73	24.14	23.40	−7.5
XY UMa	EB	0.479	9.7	1.2	G2 V + dK 5	1.09	0.81	1.90	1.15	0.96	+4.5
TZ Lyr	EB	0.528	9.4	0.95	K0 + K9	0.89	0.61	1.50	0.99	1.25	+4.8
UV Leo	EA	0.600	8.8	0.66	G0 V + G2 V	1.02	0.95	1.97	1.12	1.11	+3.7
CN Lac	EB	0.637	11.7	0.5	dG 3 +	1.00	0.57	1.57	1.09	0.87	+4.8
YY Gem	EA	0.814	9.1	0.56	M0.5 Ve + M0.5 Ve	0.57	0.57	1.14	0.60	0.60	+8.4
UV Psc	EB	0.861	8.7	1.0	dG5	1.05	0.81	1.86	1.15	1.15	+4.0
R CMa	EA	1.136	5.70	0.64	F IV + K1	1.19	0.19	1.38	1.59	1.25	+3.5
CM Dra	EA	1.268	12.87	0.78	dM3e + dM3	0.24	0.22	0.46	0.24	0.22	+12.3

from the EW stars, whose masses are often small. Amongst these, there are two red dwarfs that we mentioned in the previous chapter, YY Gem and CM Dra, the latter having the smallest known mass for eclipsing binaries (0.46 sun).

Chapter 2

Eclipsing Binaries:
Particular Phenomena

In the previous chapter, we described the various classes of eclipsing binaries and showed the importance of them in the determination of some physical parameters. We are now going to deal with particular phenomena that arise in certain of these stars. These may be mechanical, such as the variation of period or the displacement of the line of apsides; or they may be physical, in cases when one of the the components is a variable.

VARIATIONS OF PERIOD

For a long time, it was thought that the periods of eclipsing binaries were constant; it is now known that none of them are. The observed variations are small, but they are nonetheless significant through their cumulative effect, as we have already mentioned in the case of the Cepheids. The period of Algol, for example, seemed constant at 2.86731 days for a century. In 1882, there was an abrupt change and since then a continuous increase. At present it is 2.86739 days.

For some stars, the change is continuous. This is so in the case of SW Lac, whose period increased from 0.3207214 day in 1813 to 0.3207281 day at present. It is true that a change of 6.7 millionths of a day per period appears infinitesimal, but over 40 years it represents a difference of 0.3 day, almost equal to the period.

For i Boo, the period changed from 0.26781227 day in 1958 to 0.26781491 day today. For W UMa, the period increased from 1968 to 1976 and has since decreased: this is shown in the O–C diagram of Figure 95.

In some cases, the variation is abrupt. Thus, the period of U Cep has occasionally suffered large changes, one of which is shown in Figure 96.

These variations are found in all types of eclipsing binary, although the EW class appears to be the most susceptible to them. Two possible causes are as follows: in some systems there is a third star perturbing the orbit; Algol, for instance, is in reality a triple star system; and in close binaries,

188

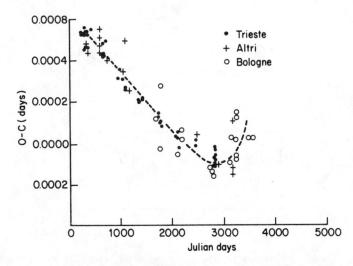

Figure 95. Variation in the period of W Ursae Majoris. (*From L. Baldinelli and S. Ghedini,* Inf. Bulletin of Variable Stars, *No. 1480, 1978. Reproduced by permission of the Konkoly Observatory, Budapest*)

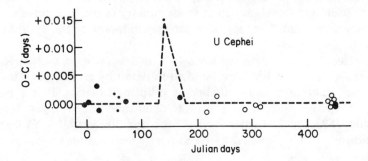

Figure 96. Abrupt variation in the period of U Cephei. (*From D. S. Hall,* Inf. Bulletin of Variable Stars, *No. 847, 1973. Reproduced by permission of the Konkoly Observatory, Budapest*)

another effect occurs, viz. the transfer of mass from one component to the other, which we shall discuss later; this transfer alters the mass ratio and thus modifies the orbit.

ECCENTRIC ORBITS

Some binaries have highly elliptical orbits, a feature that is revealed in their light curves. The effect is illustrated by Figure 97; because of the ellipticity, the distance between components A and B varies a great deal and hence the orbital velocity of B varies from one point to another. It is a maximum at the periapsis and a minimum at the apoapsis.

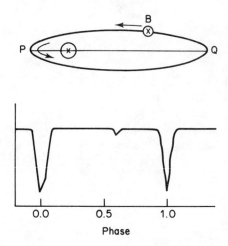

Figure 97. A binary with a highly eccentric orbit. The line PQ is the line of apsides. The lower diagram shows the displacement of the secondary minima

This variation produces a displacement of what is called the line of apsides, that is, the line PQ joining the periapsis and apoapsis. The major axis in fact turns on itself and this shows up as an asymmetry in the light curve since the secondary minimum does not occur at the midpoint of the cycle between two principal minima.

This effect has been detected since 1937 in several stars, such as RZ Cas, TW Dra, RT Per and RV Mon (studied in detail by H. Shapley). It is now known to occur in a large number of eclipsing binaries. It is a periodic effect, but the period is often difficult to determine since it is very long: 248 years for V523 Sgr, 284 years for RV Mon and 300 years for YY Sgr. One exception is GL Car, studied by H. Swope, which has an apsidal period of 25 years.

MASS TRANSFER

We have mentioned the question of mass transfer from one star to another several times, but we must now return to it once again in more detail.

Consider a binary system A and B. The Lagrangian point L is the point at which the attractive forces of the two components are equal and opposite. If one of the stars occupies a volume sufficient to reach the Lagrangian point, the surface of the volume occupied is called the Roche surface (after the French mathematician who established this theory) and the volume itself is called the Roche lobe.

Consider the case in which A is a star in the later stages of development, a red giant say, with a large volume for a relatively small mass. If it completely fills its Roche lobe, the material further from the centre than the point L will be attracted by B and will fall on to the latter's surface: there is mass transfer from A to B.

With close binaries, three cases have to be considered. In the first, neither component reaches its Roche lobe. Nothing therefore happens, except a certain amount of deformation of the stars due to their mutual attraction, which causes them to become pear-shaped. Such a system, composed of young stars in the early stages of evolution, is called a 'detached binary.' In the second case, component A reaches or exceeds its Roche lobe, while B remains well inside it. There is then transfer from A to B; such a system, where the more massive component evolves more rapidly than the other, is called a 'semi-detached binary.' Finally, we have the case in which both stars have reached their Roche lobes, when there is reciprocal mass transfer from A to B and from B to A, with the formation of a common atmospheric envelope. This is said to be a 'contact binary' and is shown diagrammatically in Figure 98.

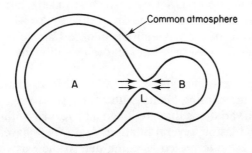

Figure 98.   Contact binaries. The two stars filling their Roche lobes are in contact at the Lagrangian point L. There is a reciprocal transfer of material and a common atmospheric envelope is formed

The mass transfers produce complex physical phenomena that are outside the scope of this book. We only mention one seemingly obvious fact: the transfer changes the mass of both components and thus their mass ratio. This produces changes in the orbit and the period, with which we have already dealt. The period increases if it is the more massive star that loses mass, and decreases if the loss is suffered by the less massive component.

Mass transfer also explains something that might astonish the reader when examining the tables in the last chapter: in many cases, the principal star is a dwarf (class V) whereas the secondary star is a giant or subgiant. Thus, if we take the example of Algol, we see that the principal component has a spectral type B7 V, an absolute magnitude of -0.6 and a mass of 3.7 suns. The secondary is a subgiant (G8 IV) of absolute magnitude 3.5 (brighter than the Sun) and of mass only 0.8 sun. This apparent anomaly is due to mass transfer; initially, the subgiant was more massive, but it lost an appreciable proportion of its mass to the other smaller star.

The transfer may be rapid; C. de Loore and his collaborators took an example of a binary consisting initially of two large stars A and B with masses of 20 and 8 solar masses. Calculation shows that the time between the beginning of mass transfer and its end is only 20 000 years, which is very short on an astronomical scale. At the end of that time, mass transfer had reduced A to 5.4 solar masses, while B had reached a massive 22.6 solar masses.

Mass transfer also produces other effects, the major one being the following. If one of the components is a very dense star (white dwarf or pulsar), material falling on to it at a high velocity raises its temperature considerably, which causes it to emit X-rays. We shall see later that a large number of close binaries are indeed important X-ray sources.

BINARIES WITH INTRINSIC VARIATIONS

Some eclipsing binaries have one, if not two, variable components. The variations are of all types. We have already encountered some of them: the ex-novae, like VV Pup or UX UMa, which are characterized by flickering (a permanent rapid variation) or the dwarf novae, also with flickering, apart from the explosions of the U Geminorum or Z Camelopardalis types. This sometimes makes the observation of the eclipse difficult (see Figure 70 relating to Z Cha).

AN UMa shows another type of variation. It is an ex-nova similar to V Sge and is also an eclipsing binary with a very short period of 0.0797 day, but it also shows variations that are long and irregular with an amplitude of 2 magnitudes and lasting several hundred days. These have been studied by Meinunger and are interpreted as being due to the existence of a gaseous cloud of varying opacity that surrounds the binary. The cloud occults the system and its varying opacity produces changes in the observed brightness.

We have also seen that the two least luminous known binaries, YY Gem and CM Dra, are red dwarfs with emission spectra which have small variations of the BY Draconis type and sometimes flares like UV Ceti (see Figure 84 relating to CM Dra). Flares have been observed in class EW binaries, such as i Boo (studied by O. J. Eggar) and U Peg (studied by M. Huruhata). This last case is shown in Figure 99; the flare, lasting 30 min, had an amplitude of 0.35 magnitude in the ultraviolet.

A very different but very interesting case is that of AS Cas, an Algol variable with a period of 1.3669 days, shown in Figure 100. The spectral types of the components are A3 + K. P. Tempesti has shown that the A3 component is a δ Scuti type star with a period of 0.058 day and an amplitude of 0.05 magnitude.

Red giants and supergiants often show intrinsic variations. This is so with VV Cep, a very long period binary mentioned in the previous chapter; the M component is a semi-regular variable. Its period has not been accurately determined but seems to be about 200 days for a maximum amplitude of 0.5 magnitude. Figure 101 shows observations made during the eclipse of 1956–58. The semi-regular variations can be clearly seen, being just as visible during the eclipse as outside it. Several stars similar to VV Cep are known, particularly AZ Cas and KQ Pup, also mentioned in the previous chapter.

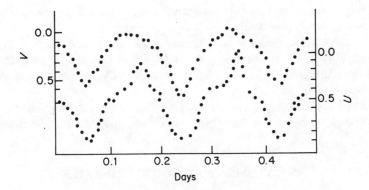

Figure 99. Light curves in *U* and *V* for U Pegasi (class EW) showing a flare in *U* that is not visible in *V*

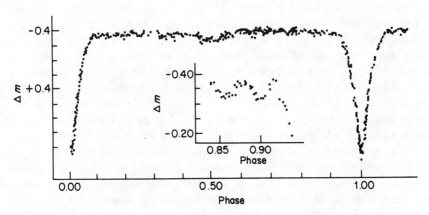

Figure 100. AS Cassiopeiae, an algol variable with δ Scuti type variations. In the inset, detail of a δ Scuti type variation. (*From P. Tempesti*, Inf. Bulletin of Variable Stars, *No. 596, 1971. Reproduced by permission of the Konkoly Observatory, Budapest*)

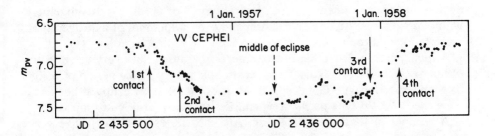

Figure 101. VV Cephei during the minimum of 1956–58, from observations of G. Larsson-Lander

We might also mention ε Aur, another long-period binary with irregular variations of small amplitude.

RX Cas, a class EB binary, is different; it has a period of 32.320 days and the spectral types of the components are G3 III and A5 III. The A5 star has a variation of unknown type, with a long period (513 days) and an amplitude of 0.4 magnitude.

We shall not quote any more examples since we are now going on to describe another group of variable binaries, of which RS CVn is the prototype.

RS CANUM VENATICORUM BINARIES

In 1931, Sitterley noticed a distortion in the maximum of the light curve of RS CVn which was of such a form that it could not be attributed to any known phenomenon such as limb darkening or ellipticity of the components. In 1946, F. B. Wood found the same effect in AR Lac and this was confirmed the following year by photoelectric observations of G. Kron. It then began to be realized that a new type of variation was being seen, but it has only been in the last decade that observations made at all wavelengths have enabled the mechanism to be understood. From that time on, RS CVn and similar stars were considered to form a special group of binaries.

RS CVn is an Algol variable with a period of 4.798 days. It consists of a subgiant of spectral type K0 IV and an F4 V dwarf, which shows intense emission lines of ionized calcium outside eclipses. Photoelectric observations clearly reveal the oscillation during the maximum, as can be seen in Figure 102. Its amplitude varies between 0.05 and 0.2 magnitude and it does not always occur at the same point in the curve. D. Hall has shown that its displacement in phase is periodic, the period being 10 years. Its amplitude also varies periodically.

Observations made with radio telescopes have revealed the existence of very intense radio flares; in addition, satellite observations have shown that RV CVn is a fairly intense X-ray source, also with abrupt flares.

At least 40 of these stars are now known, and details of a few are given in Table 49. It will be noticed that they all have one point in common: one or both of the components are of spectral type F or G (dwarfs or subgiants) with emission lines of ionized calcium and even sometimes of hydrogen that are often very strong.

A binary not in Table 49, but of interest because it lies very close to us (18 parsecs), is UX Ari. This is a spectroscopic binary with a period of 6.438 days and spectral type G5 V + K0 IV. It does not show eclipses, but does have the RV CVn type oscillations and is also a powerful X-ray source.

A number of these stars (RT Lac, AR Lac, SZ Psc, TY Pyx) have very strong radio emissions. Their flares release an amount of radio-wave energy that is a million times greater than that of solar eruptions; in February 1978, the binary V711 Tau even had an enormous radio flare that was

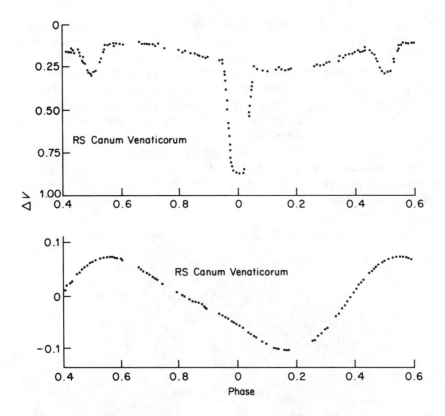

Figure 102.   RS Canum Venaticorum. The upper diagram shows the distortion in the light curve, the lower diagram the detail of the variation over a single cycle. *(From M. Zeilik, D. S. Hall, P. A. Feldman and F. Walter, Sky and Telescope, **57**, 132, 1979. Reproduced by permission of Sky Publishing Corp)*

Table 49.   Binaries of the RS CVn group. The spectra marked with an asterisk show emission lines of ionized calcium

Designation	Type	P (days)	V_{max}	ΔV	Spectral type
AR Lac	EA	1.983	6.11	0.66	G5 IV* + K0 IV*
SZ Psc	EA	2.996	8.5	1.0	K1 IV + F8 V*
Z Her	EA	3.996	7.26	0.8	K0 IV + F2 IV V*
WW Dra	EA	4.629	8.29	1.20	G2 IV + K9 IV
RS CVn	EA	4.798	8.38	1.05	K0 IV + F4 V*
RT Lac	EB	5.074	8.84	1.05	G9 IV* + K1 IV*
VV Mon	EA	6.051	9.4	0.8	dG0* + K5 IV
RW UMa	EA	7.328	10.3	1.6	F9 V* + G9 V
SS Boo	EA	7.606	10.3	1.0	dG5* + dG8
RZ Eri	EA	39.283	7.79	0.92	A5m + G8 IV

100 million times more intense than the chromospheric eruptions of our own Sun! These stars are also X-ray sources, and their X-ray flares are sometimes 10 000 times stronger than those from the Sun.

The origin of these effects seems to be the same as those of what are called the 'active regions' of the Sun, that is, of coronal prominences formed as shown diagrammatically in Figure 103. These are ejections of hot gases that are undergoing violent motion but cannot escape since they are trapped by magnetic fields. Considerable heating occurs, which causes the generation of X-rays. This is undoubtedly the same phenomenon as that in the RS Canum Venaticorum stars, but the intensity there is much greater. A gigantic luminous spot is formed and its apparent brightness varies with the orbital motion of the binary.

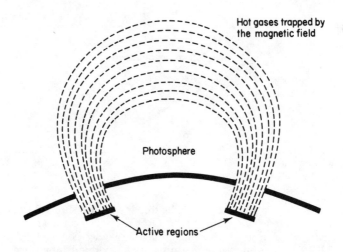

Figure 103.   A coronal prominence; the plasma emitted by the active regions is trapped by the magnetic field and cannot escape

There is no doubt that the years to come will bring more valuable information about the behaviour of stellar coronae, particularly in close binaries where the effects will clearly be amplified by the interaction between the two components.

Part 5

PECULIAR VARIABLE STARS

This final part gives an account of stars that defy classification with any of the large groups of variables already described.

The first chapter, on 'Miscellaneous galactic variables,' begins with the well known group of R Coronae Borealis stars, goes on to deal with several stars that are as strange as they are unclassifiable and ends with two extraordinary objects, V1343 Aquilae and FG Sagittae.

The second chapter, 'The new variable stars,' is devoted to those extremely dense bodies known as pulsars and to various X-ray emitters, binaries whose companion is a neutron star or a black hole, some curious X-ray novae and X-ray flare stars.

The last chapter, 'Extragalactic monsters,' is concerned with some extraordinary galaxies having variable nuclei and with the enormous bodies responsible for their luminous variations.

Chapter 1

Miscellaneous Galactic Variables

R CORONAE BOREALIS STARS

R Coronae Borealis variables form a numerically small but distinctive group of stars. They cannot be considered either as pulsating variables, as eruptive variables or as eclipsing binaries, so that they clearly form a separate class. Their normal state is at the maximum, but at irregular and generally widely spaced intervals, the brightness rapidly decreases by 4, 6 or 8 magnitudes. The reversion to the normal state sometimes occurs in a few weeks but can also take several years.

The prototype star, R Coronae Borealis, was discovered in 1795 by Pigott. It is normally of magnitude 6 with small fluctuations, but at its minima it can fall to below magnitude 14. The decrease is rapid (1 or 2 months) but the rise can take a long time. The most remarkable period lasted from 1864 to 1875; during these 11 years, R Coronae Borealis did not return to its maximum but underwent several large fluctuations. Figure 104 shows the light curve established by L. Campbell and M. W. Mayall, who used nearly 100 000 old and new observations in order to reconstruct it. Figure 105 shows some minima of R CrB and SU Tau in more detail.

The General Catalogue lists 32 stars that could be of this type, but an examination of the luminous characteristics and the spectra leads to the final conclusion that only 20 can be retained in the group, although the prototype R Coronae Borealis is undoubtedly one of them. These stars are therefore fairly rare and it is known that several of them do not belong to the Galaxy; W Men, for example, is in the Large Magellanic Cloud. Table 50 gives the properties of eight of these stars; RY Sgr and SU Tau are the most thoroughly studied of this group, apart from R CrB itself.

These stars are supergiants with G spectra and absolute magnitude around −6. Their spectra, which have been subjected to considerable study, are normally rich in carbon (sometimes denoted by the letter C) and poor in hydrogen. At the minimum, there are several metallic lines in emission (neutral sodium, ionized calcium, titanium and iron) but there are absorption lines

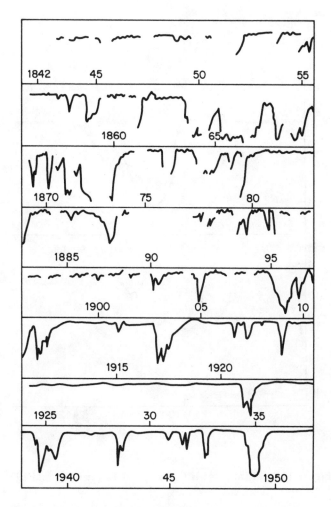

Figure 104.   Light curve of R CrB from 1843 to 1951 according to L. Campbell
and M. W. Mayall

of hydrogen. In addition, and this is an important point, the radial velocity
varies very little between minima and maxima, which rules out the possibility
that matter is being ejected.

In the last few years, a semi-periodic pulsation has been shown to exist in
some of these stars, but only of small amplitude (0.1–0.3 magnitude). The
'period' is 38.6 days for RY Sgr, 45 days for R CrB, 42 days for UW Cen and
38 days for GU Sgr. This pulsation is independent of the main variation, as
can be seen from Figure 106.

These stars generally have a small colour index; for R CrB, the $B-V$ index
is only $+0.60$, slightly lower than that of the Sun ($+0.65$). On the other
hand, observations in the infrared show that they are much brighter in that

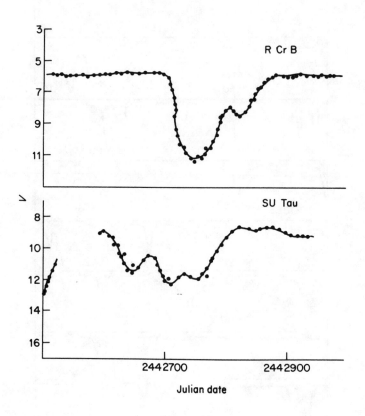

Figure 105.   Minima of R CrB and SU Tau in 1975 and 1976

Table 50.   Some R Coronae Borealis stars

Designation	V_{max}	V_{max}	Spectra
S Aps	9.79	15.2	C3
XX Cam	7.35	9.8	G1 Ie
UV Cas	10.65	15.0	
R CrB	5.60	14.8	G0 Iep
W Men	13.90	16.0	
RY Sgr	6.5	14.0	G1 Iep
SU Tau	9.77	16.0	G0 ep
RS Tel	8.5	<13.0	C5

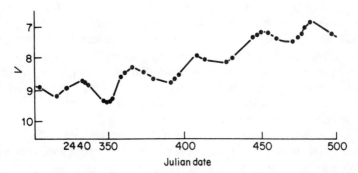

Figure 106. Periodic variation of RY Sgr during the minimum of 1969

part of the spectrum than in the visible region. For example, RY Sgr, which has a visual magnitude of 6.5 at the maximum, reaches a magnitude of −0.4 in the far-infrared; here it is brighter than Vega.

The interpretation of all these observations runs into difficulties. The stars are unusually rich in carbon and it is thought that this element plays a similar role in the star's atmosphere to that of water vapour in the atmosphere of the Earth, condensing to form clouds that obscure the sky unless they disappear in the form of rain.

The carbon circulating in the star's atmosphere condenses in the upper regions in the form of fine grains of graphite. During their fall towards the surface under the influence of gravity, these 'clouds of soot' become denser, and end up by forming a thick envelope that obscures the star and cuts out a large part of its radiation. However, when the clouds get nearer still to the surface, their temperature rises and the carbon again becomes gaseous by sublimation. The atmosphere is clear once more and everything is in order ... until the next cycle starts.

Although this model appears to be more or less satisfactory, there are still theoretical problems to be resolved, particularly that of the mechanism by which the carbon is ejected into the upper atmosphere.

SOME UNIQUE STARS

About 40 variables are classed by the General Catalogue as 'unique variables.' Some of them are in fact almost classifiable with known types, such as spectroscopic binaries with a variable component and stars possessing a bright spot. However, some of them resist all attempts at classification and these present problems to astronomers that are by no means all resolved. We cannot describe all the stars in this category, but have chosen a just few of them and try to give for each one at least a partial explanation of its behaviour.

First, there is V389 Cyg. This star oscillates between magnitudes 5.50 and 5.69 with two periods: the first, P_1, is 1.12912 day with an amplitude of

0.19 magnitude; the second, P_2, is 1.19328 day with an amplitude of 0.05 magnitude. However, these periods do not superpose as they would do if it were a normal pulsating variable. For a time which can be anything from 1 to 3 weeks, the star varies with period P_1, and after that it varies with P_2. Sometimes, between the two modes of oscillation, it stops varying at all!

Not only that, but the radial velocity also varies with yet a third period, $P_3 = 3.31322$ days, which is not a multiple of either of the others. The maximum radial velocity sometimes coincides with the minimum brightness. This is almost what happens with the Cepheids, but V389 Cyg has a B8 spectrum which is not at all that of a Cepheid. Once again, the observed facts have not been satisfactorily explained.

The variable AE Aqr is completely different. This is a binary consisting of a red subgiant with an emission spectrum K5 IVe and a red subdwarf sdB. It has been well studied by A. H. Joy and is a spectroscopic binary of period 0.4116 day. Photoelectric observations by F. Lenouvel and J. Daguillon have shown some odd photometric variations: flares are observed similar to those in UV Ceti stars but with superposed rapid irregular variations, as in dwarf novae and ex-novae. In addition, the flares are stronger when the star is brighter, as can be seen from Figure 107.

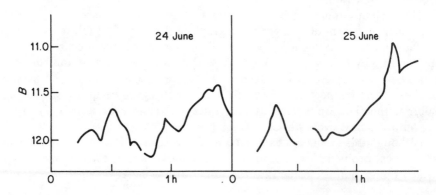

Figure 107.   Flare in AE Aquarii, from observations of F. Lenouvel and J. Daguillon

The red component is undoubtedly responsible for the flares and the B star for the flickering, but the fact that there is a correlation between the flares and the flickering shows that the two activities are closely connected. How? For the moment, nobody knows.

V348 Sgr is no less curious. From 1907 to 1927, this star varied between magnitudes 10.6 and 17, and its light curve was unhesitatingly classed among the R Coronae Borealis stars. Since 1927, the amplitude has been much smaller and the variations have become semi-periodic with a cycle of about 200 days. Its spectrum is very odd: at maximum brightness, only emission lines can be seen, with neutral helium, ionized oxygen and ionized carbon, but no hydrogen lines. At the minimum, there is still oxygen and now hydrogen in emission also: carbon, on the other hand, has disappeared. The star is also surrounded by a faint nebulosity. According to Herbig, it is probable that the luminous variations are due to the ejection of an opaque envelope. However, it is still impossible to understand why the carbon and hydrogen appear and disappear alternately.

V605 Aql is different again. It is normally very faint, with photographic magnitude 20, but it had a maximum accompanied by large fluctuations that took it to magnitude 11 between 1918 and 1924. Some observers have reported that its light curve is similar to that of the slow supernova in NGC 1058, which we have already mentioned. V605 Aql, however, is not a supernova for in 1921, at the time of its maximum, K. Lundmark found that it had a carbonized spectrum similar to that of R Coronae Borealis! In addition, it was surrounded by a small planetary nebula, A58. It is therefore rather to be classed among the novoids, but further observation needs to be carried out.

OTHER CURIOUS STARS

We have several times mentioned binaries formed from a red giant and a hot but not very bright star, first in connection with certain long-period variables such as R Aquarii, and then with symbiotic stars of the Z Andromedae type. We now describe some objects that are still more curious.

First there is V1016 Cygni: between 1920 and 1963, this star behaved like a long-period variable, oscillating between magnitudes 15.0 and 17.5 with a period of 450 days. From 1964, its brightness began to increase and in 1972 it reached magnitude 10.3. Since then, this has been maintained, except for small fluctuations. From 1965, its spectrum showed extremely strong emission lines and in 1974 a small planetary nebula was seen to appear and increase rapidly in size. We may add that, although the Mira-type variations disappeared in the visible region, they persisted in the infrared and the period was unchanged at 450 days.

The second case is that of V1329 Cyg, a star with an M5 spectrum. Between 1891 and 1964, it varied irregularly between magnitudes 14 and 16; since 1965, the brightness has increased to 11.5 and superposed on the M spectrum there is now a planetary nebular spectrum with very strong emission. In addition, V1329 Cygni displays eclipses with a period of 959 days and an amplitude of 2 magnitudes.

The last case is that of HM Sge. This star oscillated irregularly between 15 and 16 from 1950 to 1974, but in 1975 it increased in a few months to magnitude 10.5 and at the same time a planetary nebular spectrum appeared here also.

Hence in all these three cases, and perhaps in several other less well observed ones, the appearance of a planetary nebula has been witnessed. The term 'protoplanetary nebulae' is used to designate these systems, which are of great interest since they are intermediate between red variables and planetary nebulae and could possibly provide us with valuable information about the final stages in the evolution of red giants.

Before leaving the question of planetary nebulae, we describe very briefly another curious object, the central star in the planetary nebula NGC 2346. In 1978, R. Mendes discovered that this was a spectroscopic binary with a period of about 16 days. At the beginning of 1982, L. Kohoutek detected eclipses with such an amplitude (2.5 magnitudes) that it left no doubt about the reality of the phenomenon. However, an examination of photographic plates taken since 1899 shows that eclipses appeared only in November 1981; before that, the brightness did not vary. Observations made since 1983 both in the infrared and with various colours of the visible region show a continuous modification of the light curve which led A. Acker to think that the eclipses were in fact due to the passage of a cloud of material with a highly opaque central band in front of the spectroscopic binary. The phenomenon now seems to be gradually disappearing; at the end of 1984, the amplitude had become less than 1 magnitude and if Acker's theory is correct, the eclipses will completely disappear in 1986.

A ROTATING BEACON: SS 433 (V1343 AQUILAE)

This star was discovered in 1966 by C. Stephenson and N. Sanduleak and catalogued by them under the number SS433 as an object showing very strong emission lines.

However, it was only in 1978 that the work of D. Clark and P. Murdin and of B. Margon showed that this object is very exceptional. The emission lines are subjected *simultaneously* to a shift towards the red and the violet: in other words, the same lines appear twice in the spectrum, or occasionally three times, since they may be found in their normal as well as their shifted positions! Moreover, the shift is enormous; observations made by Margon and by F. Ciatti at the Asiago Observatory show that the radial velocities vary between $-35\,000$ and $+50\,000$ km/s! The motion thus changes over a total range of 85 000 km/s, an appreciable fraction of the speed of light. It is also a periodic motion with a period of about 164 days.

We should make it clear at once that this star is well inside our Galaxy. It is 3000 parsecs from us and its absolute magnitude is close to -3.5. It is a binary system with a period of rotation of 13.1 days and formed from a hot luminous O spectrum or Wolf–Rayet star and a companion that is massive (2 or more solar masses) but small, probably a neutron star. This companion is surrounded, as is generally the case, by a gaseous disc produced from material detached from the principal star.

The object is at the centre of a radio source W50, and satellite observations have shown that it is also a powerful X-ray source. This originally led to the idea that the radio source could be a supernova remnant, but observations made in the last few years using large interferometers show that the star and the radio source are linked by thin filaments. It is now thought that W50 is a large gaseous envelope in which the stellar system is immersed.

SS433 is also a variable (V1343 Aql) with an amplitude of about 1 magnitude. For a long time, there was some uncertainty about the nature of the variation; observations made by D. Overbye in 1979 and by Cherepaschuk in 1979 and 1980 show that it is in fact an eclipsing binary with a period of 13.09 days. However, the eclipse is produced by the occultation of the gaseous disc surrounding the neutron star by the principal star. The amplitude varies from one minimum to the next, but the period stays the same since it is the period of rotation of the pair.

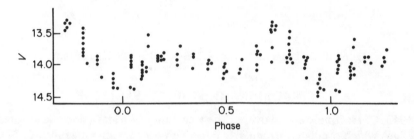

Figure 108. Light curve of V1343 Aql (adapted from Cherepaschuk)

What sort of mechanism can possibly produce such a large shift in the spectral lines? Since 1978, several theories have been put forward but none can be considered final. The most likely of these is illustrated in Figure 109, in which the neutron star is emitting two beams of very hot gas by a mechanism not yet very well understood, but at a considerable velocity of about 80 000 km/s. It can be seen in the diagram that the beams make an angle of about 17° with respect to the star's axis so that, owing to precession, they revolve about this axis in 164 days. Because terrestrial observers themselves are placed at an angle of 78° with respect to this axis, there is an asymmetry. This is the explanation of the inequality of the velocities of approach and recession. At any moment, there is one beam approaching the observer while the other recedes, and this produces the simultaneous shifts towards the red and violet.

Another interesting observation has been made by G. Collins and G. Newson: in less than 3 years, the rotation period of the beams has fallen from 167 to 162 days (164 days is a mean value). This reduction is very rapid, viz. 0.11 day per day. Such a deceleration is inexplicable but in any case it shows that the phenomenon we observe is destined to have a short life. What will happen then? For the moment, it is impossible to say.

S433 is an exceptional object that is currently fascinating a number of astronomical groups. So far, there is no other star like it.

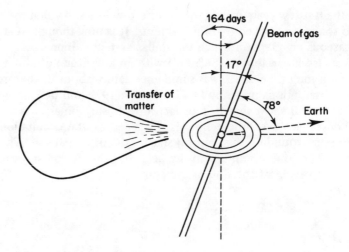

Figure 109.   Diagram of the V1343 Aquilae (SS 433) system

AN EXTRAORDINARY STAR: FG SAGITTAE

In 1943, C. Hoffmeister discovered the variability of a star (since designated
FG Sge) which oscillated irregularly between photographic magnitudes 11.0
and 11.5. It was observed once more in 1960 by G. Richter, who found
that it varied between magnitudes 10.0 and 10.5. He then looked for it on
old photographic plates from the observatories at Harvard, Sonneborg and
Moscow and saw that, in addition to the irregular fluctuations, the brightness
had been continually increasing from photographic magnitude 14 in 1894
to 10 in 1960 (Figure 110). The increase continued until 1967, when the star

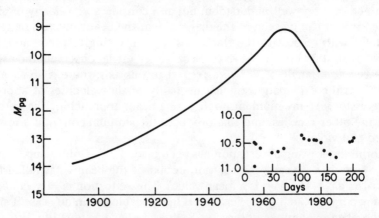

Figure 110.   Average light curve of FG Sagittae from 1894 to 1980. Inset are de-
tailed observations of J. Jurcsic and L. Szabados (Inf. Bulletin of Variable Stars,
No. 1722, 1979) showing the period of 108 days. (*Reproduced by permission of the
Konkoly Observatory, Budapest*)

reached a magnitude of 9.5, but since then there has been a decline to 10.8 in 1979. G. Herbig and K. Henize have also indicated that FG Sge is the central star of a planetary nebula.

As FG Sge became more interesting, it was followed with greater continuity, and revealed surprise after surprise. From 1894 to 1962, the star maintained its irregular fluctuations with the same amplitude of 0.5 magnitude. In 1962, however, while still keeping the same amplitude, the variations became *periodic*. FG Sge then became a pulsating variable resembling the Cδs Cepheids! The period was first found to be 15 days, and then in 1964, 22 days. Observations since then reveal that the period is increasing very rapidly, reaching 108 days in 1979! Figure 111 shows that the increase with time is linear at 4.9 days per year.

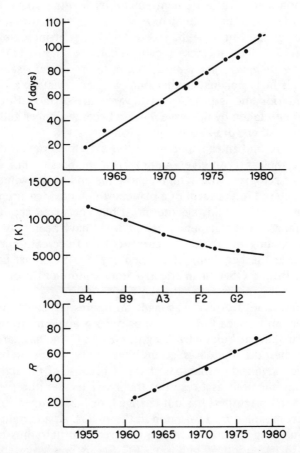

Figure 111.   Variations in some physical properties of FG Sagittae (after J. Sauval, *Ciel et Terre*, **96**, 371, 1980). Top: increase in the period. Centre: variation of temperature and spectral type. Bottom: increase in the radius. (*Reproduced by permission of Ciel et Terre*)

Studies of the spectrum reveal still more surprising results: in 1955, FG Sge was of spectral type B4 I, corresponding to an effective temperature of 12 300 K. Observations in 1967 by G. Herbig and A. Boyarchuk showed that the spectrum 12 years later was of type A5 I, corresponding to 8300 K. In other words, *the star had cooled by 4000 K in 12 years*, more than 300 K per year. And this has continued: in 1970, the spectrum was F2, and currently it is G6, corresponding to a temperature of around 5000 K. Figure 111 shows that this variation is also linear.

In 1974, Langer, Kraft and Anderson announced a still more astonishing result: between 1965 and 1972, the atmospheric concentration of certain rare heavy elements such as barium, lanthanum, cerium and zirconium had become *25 times greater*, whereas light elements had kept their normal concentrations! Since 1972, there has been no more variation.

Precise photometric measurements made by Whitney have enabled the diameter of FG Sge to be calculated. Another surprise: from a diameter of 20 times that of the Sun in 1962, there had been an increase to 75 solar diameters in 1977. This increase (again linear, see Figure 111) was at the rate of *3 solar diameters per year*, or 4.2 million km per year (11 500 km per day!). This fact explains another one: we know that pulsation starts in the stellar interior and is propagated towards the surface. But as the star expands, the path taken by the wave motion becomes longer and longer; the pulsation period therefore also becomes longer.

It is not easy to find an explanation for this extremely rapid evolution, but we shall nevertheless try to give a summary of the possibilities. FG Sagittae was probably a red giant which, on reaching the end of its evolution, ejected some of its material in the form of a planetary nebula. Measurements of the radial velocity show that this ejection must have occurred about 6000 years ago, and the mass of the planetary nebula must have been between 0.2 and 0.6 sun. The cooling was rapid: the current rate of decrease in temperature suggests that, at the beginning of the century, FG Sge must have been of spectral type around O5 with an effective temperature of the order of 40 000 K.

In this connection, J. Sauval has made an interesting comment: if a bolometric correction is made to the visual or photographic magnitudes, then it becomes clear that the bolometric magnitude of FG Sge has hardly changed from 1894 (when the first observations were made) to the present. It is the photographic magnitude which has changed because of the star's considerable alteration in colour as a result of the lower temperature.

Since the total brightness has not changed, or has changed very little, the cooling is accompanied by a variation in the radius. At a certain instant, the star starts pulsating, but the continuing increase in its radius brings about an increase in its period of pulsation. FG Sge is now counted among the Cepheids of Population II or the RV Tauri stars, as much because of its period as its absolute luminosity and temperature. On applying a well established relationship between the period and the mean density, a very small mass of about 0.4 solar mass is found for FG Sge. This value undoubtedly needs to be confirmed, but it is believed in any case that the mass of this star is at most equal to that of the Sun, which is very small for such a bright star ($M_{bol} = -4.5$).

The rest of the explanation must be sought in the theory of the evolution of stars of low mass. When a star has exhausted its hydrogen, it starts to grow and cool and then moves towards the region of the H–R diagram occupied by red giants. At the same time, it turns to another nuclear reaction: the transformation of helium to carbon and oxygen. However, this reaction does not proceed as smoothly as hydrogen fusion. It occurs in the form of flashes more or less spaced out in time, and these have the tendency to destabilize the star. One effect of the flashes is to drive away a large part of its exterior envelope to form a planetary nebula surrounding a small very hot star called the nucleus of the nebula.

This event happened in FG Sge about 6000 years ago. It remained at this stage for a time, while the nucleus was slowly cooling. In less than a century it has evolved with surprising rapidity, since from 1955 to 1983 it travelled across two thirds of the H–R diagram from left to right (from B4 to K0). If it carried on at its present rate, it would reach the stage of a late M spectrum in 20 years. What would happen to it then? It is difficult to say, since the evolution depends mainly on the mass and that is not known. There are two possibilities, however. The first is the following: with its enormous radiation for such a small mass and exhausted by its ejections of matter, the star becomes subject to gravitational contraction; it collapses and passes directly to the red giant or white dwarf stage. The second possibility was suggested by Paczynski: a final series of helium flashes sends the star back once more to the region of hot stars, but it does not stay there; it descends the H–R diagram and moves towards the white dwarf region. In either case, it winds up at the same destination: transformed to a white dwarf, the end awaiting all stars of small mass.

FG Sagittae is a fascinating object: firstly, because it enables us to observe for the first time the actual synthesis of heavy elements predicted by theoreticians; secondly, because it is exceptional that a star should evolve so appreciably in a few decades; lastly, because it shows us that the nucleus of a planetary nebula does not necessarily travel directly to the white dwarf region as was believed until now, but can pass through a second red giant stage. This is a new fact that needs to be explained by theoreticians.

In my view, we have had a lot of luck in being able to observe these phenomena, since the probability of following a phase of development that lasts less than a century is very small. Is this to say that such a phase will never be observed again on another star? Our best chance is to watch the planetary nebulae and their nuclei—but do these nuclei necessarily pass through the FG Sagittae stage, or is this star unique? For the moment, it is impossible for us to answer that question.

Chapter 2

The New Variable Stars

The use of artificial satellites has led to the discovery of new variable stars through the development of a high-energy astronomy, in other words one based on observations made at X-ray and gamma-ray wavelengths. Radioastronomy, too, can take the credit for the discovery and study of those extraordinary stars, the pulsars.

PULSARS

The first pulsar was discovered fortuitously. In November 1967, J. Bell, using the Cambridge radiotelescope, found a radio source CP 19–19 showing pulses that were very short but repeated themselves with a period of 1.3373 s. This so amazed the Cambridge astronomical team that it was only announced several months later after being checked many times. The shortest periodic variations known at the time were of the order of 1h, which demonstrated that an entirely new type of object was being observed. Searches were then carried out for other stars similar to CP 19–19, called 'pulsars' from an abbreviation of 'pulsating stars.' The search turned out to be very fruitful: the General Catalogue of 1976 already recorded 176 of them and more than 300 are known today.

Their periods are always very short: the longest known is about 4 s and the shortest is that of a pulsar we have already mentioned, NP 05-32, the central star of the Crab Nebula, with a period of only 0.033 s. The frequency distribution of the periods shows two maxima, one at 0.7 s and the other about 1.4 s. A point to be noted is that, if a pulsar is observed at various radio frequencies, the intensity of the pulses may be different but the period stays the same. This can be seen in Figure 112 for the pulsar PSR 0329 + 54.

High-precision measurements have shown that the periods are not completely constant, however, but are slowly increasing. It is true that the increase is very small: on average it is no more than 10^{-8} s per year!

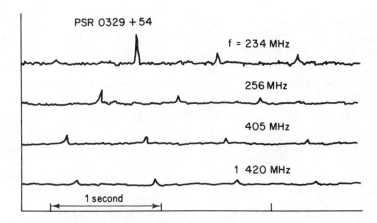

Figure 112. Variations of the pulsar PSR 0329 + 54 whose period is 0.7145186 s. Measurements were made at four different radiofrequencies from 234 to 1420 MHz, and show that the period remains the same as the frequency changes, whereas the amplitude varies

A pulsar is a neutron star and has an extremely high density. How is it formed? We have seen that after the gigantic explosion of a supernova and the ejection of a large part of its mass in gaseous form, a small nucleus of 10–20% of the initial mass remains, and this collapses upon itself. The remnant consists mainly of neutrons, heavy particles that have no electric charge and therefore do not repel each other but form a highly compact agglomerate, which is the main part of what is called a neutron star.

Pulsars are extremely small stars; work carried out from 1968 by A. Hewish and M. Ryle, and subsequently by others, has shown that the radius of a neutron star of one solar mass is no more than 10 km. This produces a fantastically high density, much greater than that of white dwarfs. To take an example, the density of NP 05-32, the Crab pulsar, relative to water is 5×10^{14}, which means that 1 cm^3 of it has a mass of 500 million tonnes! These stars rotate rapidly about their own axes, and this generates a magnetic field that is also enormous, 10^{10}–10^{12} G. Some pulsars have been observed to be X-ray sources and four of them can also be observed in the gamma-ray region. In this region, it has been noted with some surprise that there are two peaks in intensity separated by 0.4 of the total period, whereas in other regions, particularly that of radio waves, there is only one peak. For the moment, this remains a mystery.

We should add that very few pulsars radiate in the visible region also; we know of only NP 05-32 (the Crab pulsar), PSR 0833 Vela and the remnant of the 1572 supernova.

The explanation of the pulses is as follows. Because of the enormous strength of the magnetic field, the particles emitted by the pulsar are trapped by the field and can only escape through a region near the magnetic poles.

The fact that the pulses are extremely short means that this region must be very small. The rotation and magnetic axes of the star are probably not coincident (Figure 113); this is generally the case in stars just as it is with planets. The magnetic axis revolves around the rotation axis, so that the star forms a rotating beacon, similar to a lighthouse beam that sweeps through the sky. The radiation leaving along the magnetic axis is only observable if the observer is situated in the region swept by the beam, which is clearly not the case for every star. It has been calculated that only one pulsar in 30 is observable from the earth, with the rest sending their radiation in random directions.

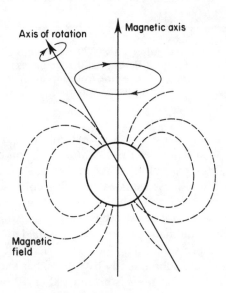

Figure 113. A pulsar

With several hundred of these stars already detected, they are clearly relatively numerous, so how do they originate? We have said that a pulsar is a supernova remnant and this has been verified in several cases. But are *all* pulsars supernova remnants? Although this is plausible, because there must have been a large number of supernovae in the time since the galaxies were first formed, it is still difficult to reply in the affirmative.

A very interesting feature has been the discovery of double pulsars. Three are known so far, the most thoroughly studied being CP 1913 + 16, with an orbital period of 0.323 day or 7 h 45 min. It is formed from two pulsars: A, with a moderately long period of 1.616 s, and B, with a very short period of 0.059 s. According to what has been said in the chapter on supernovae, the origin of such a pair can be imagined as follows. Because of the mass transfer from A to B, the A component was reduced to its helium core, had a supernova phase and then became a pulsar. The transfer then began to take place in the reverse direction from B to A. Following this, B in its turn became a supernova and was then left as a pulsar or neutron star.

Such pairs may be numerous, but they are obviously difficult to detect since both of them must send their radiation in the direction of the earth for this to be possible.

X-RAY BINARIES

In the last 10 years or so, satellite observations have revealed the existence of binary systems in which one of the components is an emitter of X-rays, often a very strong one. The first of these systems was discovered by R. Giaconni and his collaborators in 1971 using the Uhuru satellite. This is the source Cen X3 (since called V779 Cen). It exhibits an X-ray pulsation of period 4.842 s. The system also shows eclipses with a period of 2.08726 days, and during the eclipses, the pulsation stops; it is clearly the X-ray component that is eclipsed.

The mechanism of the X-ray emission is as follows. A binary such as Cen X3 is formed from a normal hot star of O or B spectrum and a neutron star. Material escapes from the normal star and flows towards its companion, either because the former occupies the whole of its Roche lobe (Figure 114) or because the 'stellar wind'* is captured in part by the neutron star. When the gas falls on to the companion, part of its kinetic energy is transformed into heat energy; in other words, it forms a very hot plasma on the neutron star and thus generates high-energy radiation, X-rays. The plasma is trapped by the very strong magnetic field of the hyperdense star in the regions around the magnetic poles and this produces periodic pulses, similar in principle to those of the pulsars.

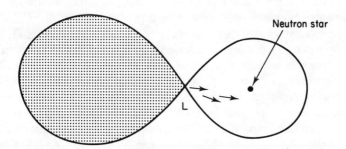

Figure 114.   An X-ray binary. Material from the normal star flows towards the neutron star through the Lagrangian point L

---

* A stellar wind (or solar wind for the Sun) consists of particles ejected by the star at sufficient velocities to escape its gravitational attraction.

Several systems of this type have been discovered, and details of five of them are given in Table 51, including their visible spectra and the X-ray power emitted in J/s. The X-ray power is always enormous, from 10^{29} to 10^{31} J/s. When we recall that the total emission from the Sun over all wavelengths is only 10^{26} J/s, we can see that the power emitted by X-ray binaries is several thousand times greater.

The importance of binary systems is that they enable the masses of the components to be determined. It can be seen that, in the last four cases in Table 51, the mass of the hyperdense companion is between 1 and 2 solar masses. Such values are compatible with theory, which predicts the limiting mass of a neutron star to be between 2.5 and 3.5 solar masses.

The last star in the table, Cyg X1 or V1357 Cyg, is very different from the others, first because its companion does not pulsate but only shows a slight ultra-rapid flickering, and secondly because it has a mass between 5 and 6 suns, and thus above the predicted theoretical limit. Some astronomers have assumed that the companion is one of the famous *black holes* so often talked about, and experimental verification of this is being sought.

Let us recall what a black hole is: if a star is even more condensed than a neutron star, the gravitational force at its surface will be so great that the escape velocity for particles leaving the star will become greater than the speed of light. In that case, *radiation cannot escape from the star*, since even a photon that travels with the speed of light does not reach the escape velocity! It follows that a black hole can receive radiation from outside but cannot emit it. By definition, *it is not directly observable*; it can only be *assumed* to exist, for example if transfer phenomena are observed.

Cygni X1 seems to be the most serious candidate for the title of 'black hole,' but it is by no means a certainty. It is true that the irregular and very rapid pulsations observed in the X-ray region (very different from the regular pulses of pulsars) lead us to believe that it is, but observations with a more sophisticated X-ray satellite than those currently in service would be needed to demonstrate the fact beyond question. Some astronomers think that the companion is more likely to be a pair of pulsars.

There are other candidates, particularly Cir X1 (BQ Circinus), but they are equally open to question. We have to admit that at the moment *there is no black hole whose existence can be demonstrated with certainty*. We leave this problem and go on to give some more details about X-ray binaries.

It has been calculated that the principal star in such a binary may lose up to 10^{-6} solar masses per year by transfer (that is, one solar mass in a million years) partly to its companion. The increase in mass of the latter must affect the rotation period appreciably and this can be seen in Cen X3. Figure 115 shows that the period of this system decreased by 0.006 s in 4 years, which is not a negligible rate; in 1000 years, the period would fall from 4.8 to 3.3 s. It can therefore be assumed that the lifetime of such a system is very short.

Another type of X-ray binary exists in which the neutron star is at least as massive as, if not more massive than, the visible component, which in this

Table 51. Some X-ray binaries of large mass

Designation	Orbital period (days)	Pulsation period (s)	m_{pg}	Spectra	Distance (kpc)	Luminosity (J/s)	Masses (Sun = 1)
Cen X3 (V779 Cen)	2.1	4.8	13.4	O6.5 II	8.0	4×10^{30}	$17 + 1.0$
Vel X1 (GP Vel)	9.0	283	6.9	B0.5 Ib	1.4	1×10^{29}	$20.5 + 1.7$
4U 1538-522 (QV Nor)	3.7	529	14.5	B0 I	7	4×10^{29}	$20 + 2.0$
SMC X1	3.9	0.7	13.3	B0 Ib	65	6×10^{31}	$16.2 + 1.0$
Cyg X1 (V1357) Cyg)	5.6	Irreg.	8.9	O9.5 Iab	2.5	2×10^{30}	$30 + 6$

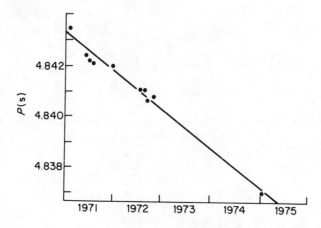

Figure 115.   Changes in the X-ray pulsation period of Cen X3 (V779 Cen) from observations made by satellites Uhuru and Ariel 5

case is a dwarf. This is so with Sco X1 (V818 Sco), which was the first X-ray source located in the sky. It is a binary with a period of 0.79 day; it has no pulsation but does show a continuous rapid flickering and sometimes some actual flares, detectable not only in the X-ray region but also in the visible region. This is a relatively close star (700 parsecs) and its X-ray luminosity is 2×10^{30} J/s.

AM Herculis is another example, and this is the prototype star of a small distinctive group. It is a very short period binary ($P = 0.129$ day or 3 h 6 min) formed from a collapsed star of white dwarf type and a red dwarf with an M4V spectrum. Such a pair resembles the dwarf novae of the SU Ursae Majoris group, but is distinguished from them by the fact that it does not exhibit the explosions typical of dwarf novae. In fact, the star normally oscillates around magnitude 13 with an amplitude of 0.5 magnitude, occasionally falling to magnitude 15. It is also distinguished by the fact that the light emitted from it is strongly polarized and that the polarization varies with the orbital period. The X-ray emission from the hot component also varies considerably with the orbital period.

At present, only a half dozen stars are known that resemble AM Herculis and all are binaries with periods between 2 and 3 h.

The source Her X1 (or HZ Herculis) is also curious and difficult to interpret. It shows a rapid X-ray pulsation of period 1.24 s and also an optical variation with a period of 1.700 days. At the same time, the intensity of the X-ray source varies regularly every 35 days: for 11 days it is a strong source and then it almost disappears for 24 days, followed by another normal period. In addition, the spectral type varies from B8 to F0 with a period of 1.700 days.

An explanation of these phenomena has been given by Wenzel and seems

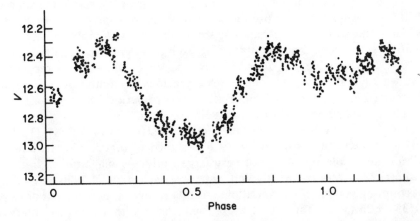

Figure 116. Eclipse of AM Herculis. (*From P. Skody and B. Browalek*, Sky and Telescope, **53**, *351, 1977. Reproduced by permission of Sky Publishing Corp*)

perfectly plausible: HZ Herculis is a binary formed from a 'normal' star with an F spectrum and a pulsar surrounded by an accretion disc consisting of material from the F star; its orbital period is 1.700 days. The pulsar is responsible for the rapid X-ray pulses (1.24 s) and the accretion disc is an intense X-ray source. The hemisphere of the F component that is facing the condensed star is strongly heated by the X-radiation and thus, when it is turned towards us, the star seems hot (B8) and the brightness is a maximum. When we can see only the non-heated hemisphere, the spectral type is F0 and the brightness decreases by nearly 2 magnitudes. As for the 35-day period, this is explained by precession—as far as we are concerned, the accretion disc is only eclipsed once every 35 days.

X-RAY NOVAE

Observation of the sky in the X-ray region of the spectrum has produced a host of surprises. Several groups of what are called transient X-ray sources have been discovered in which the emission of X-rays shows abrupt increases.

One such group is that of the 'X-ray novae.' A large and abrupt increase in X-ray emission is observed, followed by a continuous decline. The effect is accompanied by a variation of luminosity in the visible region. Thus, the object N1705-25 had an 'X-ray nova' phase in September 1977 during which the photographic magnitude changed from 21.0 to 16.5; the variations in the X-ray and visible regions were simultaneous. The X-ray energy released was enormous and at the maximum was 2500 times greater than that emitted in the visible region.

220

Another object which has turned out to be of great interest is the source A620-00. In August 1975, it produced a nova phase of appreciable amplitude and for a time was the brightest X-ray source in the sky. A visible nova phase occurred at the same time, the star rising from photographic magnitude 20.2 to 10.4. This variable was already known as V616 Mon—it is a recurrent nova we have already mentioned whose previous maximum of magnitude 12 occurred in 1917. It is also an orbital binary of period 7.8 days consisting of a collapsed star and a dwarf with a K5 spectrum.

Other recurrent X-ray nova are known. First there is Aql X1 (V1333 Aql) for which the maximum of 1978 is shown in Figure 117. Its amplitude in the visible region is not very large (only 2–3 magnitudes) but explosions occur very frequently and were observed in 1975, 1976 and 1978, not counting several minor variations. Another interesting recurrent nova is A0535 + 26 (V725 Tau); its X-ray maxima recur with an average interval of several months.

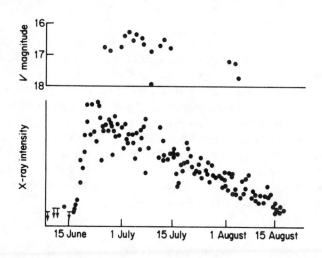

Figure 117. The 1978 maximum of the nova Aquilae X1 (V1333 Aquilae). (*From P. Charles,* Sky and Telescope, **59**, *190, 1980. Reproduced by permission of Sky Publishing Corp*)

X-ray novae have become very important because a number of 'classical' novae are not strong X-ray sources. Thus, the nova Cygni 1975 had an enormous visual amplitude but its X-ray emission was undetectable. This leads us to think that the mechanism of X-ray novae is not the same as that of the 'normal' novae, a belief reinforced by the fact that the X-ray explosions seem to occur more frequently. It must be said, however, that until now nobody has succeeded in giving a satisfactory physical explanation for the differences between the two families of novae.

X-RAY FLARE STARS

Another group of transient X-ray stars that has been discovered is that of X-ray flare sources. These flares are non-periodic, * generally lasting only a few minutes and sometimes less, as is shown in Figure 118. At the time of the maximum, the X-ray flux is 1000–10 000 times greater than the visible flux. The flares are often separated by an interval of several hours, if not several days. MXB 1730-335, shown in Figure 119, is one exception. Its very short and comparatively weak flares occur irregularly, the interval varying between 10 s and several minutes.

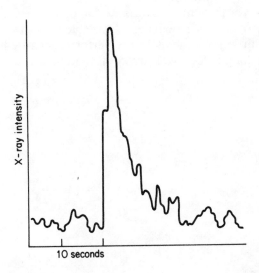

Figure 118.   Detail of an X-ray flare in NGC 6624 (arbitrary intensity scale)

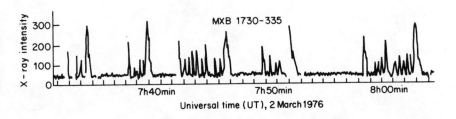

Figure 119.   Flares in MXB 1730-335 (arbitrary intensity scale). (*From H. G. Lewin, Sky and Telescope,* **54**, *364, 1977. Reproduced by permission of Sky Publishing Corp*)

---

* Some sources do have a certain periodicity, however; an example is MXB 1742-297.

An important point to realize is that, among the thirty or so flare sources currently known, most are concentrated towards the centre of the Galaxy. In addition, several have been observed in globular clusters (NGC 1851, 6441, 6624, 6712). It therefore seems that these stars belong to Population II, which makes them very different from the red dwarfs we know as flare variables. Some of these X-ray sources can be identified, such as MXB 1827-05 now called MM Ser, or MXB 1733-44 (V926 Sco); they are optically variable and have a spectrum resembling that of dwarf novae. It is thought that, here too, we have a binary containing a neutron star surrounded by a gaseous ring. The flares are probably produced by frequently repeated small explosions occurring in the ring.

Although X-ray astronomy has existed for only a few years, it has led to the discovery of several new types of variable, and this is probably only a beginning. With improvements in the performance of equipment, it will probably not be long before we learn of new and even more extraordinary discoveries.

Chapter 3

Extragalactic Monsters

Some galaxies have a central 'nucleus' that is extremely compact, very bright (emitting on its own as much light as the rest of the galaxy to which it belongs) and variable also. They are called 'active galaxies,' although this is not a very appropriate name—'galaxies with active nuclei' would be better, since they appear to be normal outside the central regions.

Active galaxies are divided into several large classes: radio galaxies, Seyfert or Markarian galaxies, BL Lacertae objects and quasars. Radio galaxies are outside the scope of this book and we shall not say anything about them, but the other types will each be briefly described before going on to deal with their variations.

Active galaxies are comparatively rare, forming less than 5% of the total, but their frequency varies with the type; there are no more than 2% of them among the spirals (generally of high luminosity), whereas they form 10% of giant elliptical galaxies.

SEYFERT GALAXIES

These objects, discovered by C. Seyfert in 1943, are galaxies with a star-like nucleus but very bright. Their spectra show very broad and very strong emission lines, and two groups can be distinguished. The first, type 1 Seyferts, are characterized by very strong broad hydrogen lines together with forbidden lines belonging to highly ionized elements such as oxygen O III that are narrower. The existence of rotational or turbulent motion reaching speeds of several thousand km/s has been noted.

Type 2 Seyferts have the opposite spectral characteristics: narrow permitted lines and broad forbidden lines. They are generally radio sources of variable intensity, unlike type 1 Seyferts, whose power in the radio region is minimal.

There are also structural differences between the two types: Seyfert 1 galaxies are generally spiral and Seyfert 2 elliptical.

The expression 'Markarian galaxies' often appears in the literature. These are objects whose spectra show high intensity in the ultraviolet and blue regions, and their name derives from the fact that an important catalogue of them was assembled by the Soviet astronomer Markarian. In fact, the term is used a little in two senses: the Markarian catalogue includes a fair number of Seyfert galaxies, in addition to BL Lacertae objects and quasars.

Seyfert galaxies are intrinsically very bright, often much more so than a normal galaxy. Work carried out in recent years has shown that Seyfert type 2 galaxies have an absolute magnitude between -18 and -20, whereas Seyfert type 1 are on average even brighter, some of them being between -20 and -24.

BL LACERTAE OBJECTS

Very few of these are known. They have very bright nuclei, but they are even smaller than those of the Seyfert galaxies. Their spectra are often continuous or show only a few weak lines, which makes them distinct from Seyfert galaxies. They are objects of enormous size and very bright indeed: their absolute magnitudes are generally between -22 and -24. A number are important radio sources.

QUASARS

These were discovered by J. Greenstein and Schmidt in 1964 and have a star-like appearance; hence their name, a contraction of the expression 'quasi-stellar objects.'

Quasars are very blue objects. Their spectra show very strong emission, particularly in the ultraviolet, but their most surprising aspect is the red shift of their spectral lines due to the Doppler effect. This shift is extremely large, so large that in some cases the quasar seems to be moving away from us at velocities that are two, three or even four times that of light! This phenomenon has been long discussed, not only because it calls into question Hubble's law (the relationship between the speed of recession of a galaxy and its distance), but also because it throws some doubt on the General Theory of Relativity. The problem is partly resolved by using a relativistic expression for the Doppler effect, but some doubt still exists as to how much of the observed red shift is produced by gravitational effects as predicted by the General Theory of Relativity and how much is due to a physical cause as yet unkown.

We do not intend to develop here the theories which have been put forward, and the reader is referred to the numerous publications on the subject. We merely record such facts about them that concern us: they are very distant objects but are extremely luminous, with absolute magnitudes that vary in most cases between -24 and -26. They are in fact the most luminous objects in the Universe.

VARIATIONS OF THE NUCLEI

We now turn to the variations that occur in the nuclei of active galaxies, beginning with the Seyfert galaxies. These vary not only in the visible, infrared and ultraviolet parts of the spectrum but also in the X-ray region. In addition, the type 2 Seyferts are variable in the radio-wave region, unlike the type 1 Seyferts.

The variations may sometimes have a periodic form, although not with a very strict period. With NGC 4151, for example, Lyutij found a period of 130 days and an amplitude of about 2 magnitudes for the visible region, but in other cases, such as that of the elliptic galaxy NGC 1275 shown in Figure 121, the variations are irregular. X-ray variations are more rapid than those in the visible region. This is illustrated for the above-mentioned galaxy NGC 4151 by Figure 120, which shows X-ray variations over a period of 12 days.

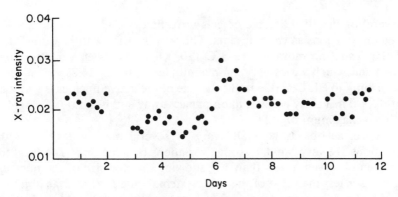

Figure 120.   Variations in X-ray flux from the Seyfert Galaxy NGC 4151 observed by the satellite OSO8 in 1977

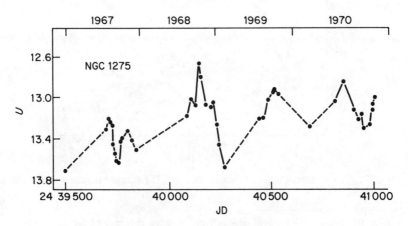

Figure 121.   Variations in U of the Seyfert Galaxy NGC 1275 from the observations of V. M. Lyutij

226

The speed of the variations is a very important point: any change in a luminous source as a whole is concerned with the propagation of a signal which must successively reach all the points in that source. This means that if a nucleus has a period of 130 days, its greatest dimension (its diameter, if it is spherical) *cannot be greater than 130 light-days*, or slightly more than a third of a light-year. On the scale of a galaxy with a diameter often exceeding 100 000 light-years, the nucleus is truly minute yet highly luminous.

The BL Lacertae objects take their name from a star discovered in 1929 by C. Hoffmeister that was first classified as an irregular variable. It was not until 1968 that J. McLeod and B. N. Andrew associated it with a known radio source, VRO 42 22 01, and it was then realized that it was in fact the nucleus of an elliptic galaxy. BL Lacertae varies from magnitude 12.9–16.6. It sometimes has a rapid rise, followed by a slow decline interspersed with oscillations, but it can also remain for several months or even a year without varying.

Several of the BL Lacertae objects were first considered as stars and received designations as variable stars. This was the case with W Com, AP Lib and BW Tau (the radio source 3 C 120, which has been very thoroughly studied and which varies between magnitudes 14.0 and 14.8).

Returning to BL Lacertae itself: the size of the active nucleus is less than 7 light-days. The galaxy is 380 million parsecs away and its absolute magnitude reaches a maximum of −23.

BL Lacertae objects generally vary by several magnitudes* and their fluctuations are very rapid. In some, variations by a factor of 2 in brightness have been observed in less than 1 h, showing that the size of the nucleus in such cases is less than that of the solar system. Some BL Lacertae objects are known to be X-ray sources; an example (Markarian 501) is shown in Figure 122.

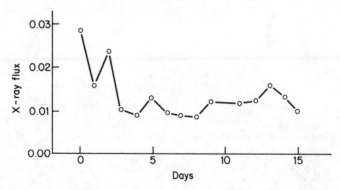

Figure 122.   Variations of X-ray flux from the BL Lacertae object Markarian 501

---

* In 1975, the object AO 0235 + 164 suffered a sharp increase of 5 magnitudes. At its maximum, the energy released was 10^{41} J/s, which made it the most luminous object in the Universe.

Several hundred quasars are now known, many of them visual variables with an amplitude that ranges from a few tenths to 2 magnitudes, while variations also occur in the radio and X-ray regions. None of the variations show much periodicity, and it should be noted that they are much slower than those of the previous groups, with some lasting for several years. This implies that the active nuclei here are far larger than those of the Seyfert galaxies or of the BL Lacertae objects. The best known of the variable quasars is 3C 273, which varies between magnitudes 12 and 13.0 but suffers long and irregular oscillations. On the whole, quasars vary less than the Seyfert galaxies or the BL Lacertae objects, but what changes there are are more rapid in the X-ray region: for example, the X-ray emission from the quasar OX 169 varies by a factor of 2 in 3 h.

THE ORIGIN OF ACTIVE NUCLEI

All these objects have one feature in common: they have a luminous source that is small in size but releases an enormous amount of energy, exceeding 10^{35} J/s for the weakest Seyfert galaxies and reaching 10^{40} J/s for certain quasars such as 3 C 273. However, if energy were to be emitted at 10^{40} J/s for 10 million years, it would need the conversion of 10 000 million suns into its energy equivalent at an efficiency of nearly 100%. It therefore seems impossible that a quasar or any other active object could keep up this rate of emission for long.

Several theories have been put forward to explain this extraordinary release of energy: the chain explosion of supernovae, gravitational contraction of supermassive stars, the collision of stars in a region of high stellar density and even the annihilation of matter with antimatter. In the end, all of these seem unlikely. More and more, it is believed that the source is a unique object, with a huge mass of the order of 10^8 solar masses (100 million suns!). Such an object can only be stable if it rotates rapidly, thus producing an enormous magnetic field.

One comment should be added here: we now know that our Galaxy also possesses a central nucleus, but one that is much smaller than those of the Seyfert galaxies, the BL Lacertae objects or the quasars. Its mass is 6–8 million solar masses and its diameter is probably no more than 1 light-hour.

To explain the existence of a supermassive object, a bold but nevertheless plausible hypothesis has been put forward: in a region of high stellar density, such as the central region of a galaxy, a star that is more massive than the others can gradually attract the stars surrounding it and absorb them, producing a considerable amount of energy. A 'hyperstar' is formed in this way whose mass may be as much as 10 000 suns. However, by its very nature, this hyperstar is unstable; at a given moment, it will collapse on itself and give rise to a hyperdense body, in this case a black hole. The enormous attraction exercised by this black hole will cause it gradually to absorb other stars and the gas surrounding it. Before being completely absorbed, these stars break up, releasing a large amount of energy. A very large black hole will thus be produced, having a mass as great as 10^8 suns. It will not itself radiate, as we know, but the material falling on to its surface does do so.

This theory, known as 'stellar cannibalism,' is supported by the fact that the dimensions of some active nuclei correspond roughly to those that a black hole of this mass would have. Such a black hole would probably have a relatively short active life (of about 10 million years) since it creates a vacuum around itself.

We are still astounded by the sheer extent of these galactic phenomena. They seem extraordinary to us, and yet observation shows that they are not at all exceptional if we take into account the large number of active nuclei in the Universe.

Appendix A

List of Constellations

Since 1930, the International Astronomical Union (IAU) has recommended astronomers to use the Latin name of constellations so as to avoid the use of designations that are often very different in different languages. Although this has been strictly followed by astronomers, it is regrettable that some authors of popular works have appeared to ignore this good advice that was given more than half a century ago! We have as a rule followed the IAU recommendation in this book but, to enable the reader to find the information quickly, this table gives the Latin name of each constellation, its genitive form, the symbol used and the English name.

Full Latin name	Genitive form	Symbol	English name
Andromeda	Andromedae	And	Andromeda
Antlia	Antlae	Ant	Air Pump
Apus	Apodis	Aps	Bird of Paradise
Aquarius	Aquarii	Aqr	Water Bearer
Aquila	Aquilae	Aql	Eagle
Ara	Arae	Ara	Altar
Aries	Arietis	Ari	Ram
Auriga	Aurigae	Aur	Charioteer
Bootes	Bootis	Boo	Herdsman
Caelum	Caeli	Cae	Chisel
Camelopardalis	Camelopardalis	Cam	Giraffe
Cancer	Cancri	Cnc	Crab
Canes Venatici	Canum Venaticorum	CVn	Hunting Dogs
Canis Major	Canis Majoris	CMa	Great Dog
Canis Minor	Canis Minoris	CMi	Little Dog
Capricornus	Capricorni	Cap	Sea Goat
Carina	Carinae	Car	Keel
Cassiopeia	Cassiopeiae	Cas	Cassiopeia
Centaurus	Centauri	Cen	Centaur
Cepheus	Cephei	Cep	Cepheus
Cetus	Ceti	Cet	Whale
Chameleon	Chameleontis	Cha	Chameleon
Circinus	Circini	Cir	Compasses
Columba	Columbae	Col	Dove
Coma Berenices	Comae Berenices	Com	Berenice's Hair
Corona Australis	Coronae Australis	CrA	Southern Crown

Full Latin name	Genitive form	Symbol	English name
Corona Borealis	Coronae Borealis	CrB	Northern Crown
Corvus	Corvi	Crv	Crow
Crater	Crateris	Crt	Cup
Crux	Crucis	Cru	Southern Cross
Cygnus	Cygni	Cyg	Swan
Delphinus	Delphini	Del	Dolphin
Dorado	Doradus	Dor	Swordfish
Draco	Draconis	Dra	Dragon
Equuleus	Equulei	Equ	Little Horse
Eridanus	Eridani	Eri	River
Fornax	Fornacis	For	Furnace
Gemini	Geminorum	Gem	Twins
Grus	Grucis	Gru	Crane
Hercules	Herculis	Her	Hercules
Horologium	Horologii	Hor	Clock
Hydra	Hydrae	Hya	Sea Serpent
Hydrus	Hydri	Hyi	Water Snake
Indus	Indi	Ind	Indian
Lacerta	Lacertae	Lac	Lizard
Leo	Leonis	Leo	Lion
Leo Minor	Leonis Minoris	LMi	Little Lion
Lepus	Leporis	Lep	Hare
Libra	Librae	Lib	Scales
Lupus	Lupi	Lup	Wolf
Lynx	Lyncis	Lyn	Lynx
Lyra	Lyrae	Lyr	Lyre
Mensa	Mensea	Men	Table (Mountain)
Microscopium	Microscopii	Mic	Microscope
Monoceros	Monocerotis	Mon	Unicorn
Musca	Muscae	Mus	Fly
Norma	Normae	Nor	Rule
Octans	Octantis	Oct	Octant
Ophiuchus	Ophiuchi	Oph	Serpent Bearer
Orion	Orionis	Ori	Orion
Pavo	Pavonis	Pav	Peacock
Pegasus	Pegasi	Peg	Pegasus
Perseus	Persei	Per	Perseus
Phoenix	Phoenicis	Phe	Phoenix
Pictor	Pictoris	Pic	Painter
Pisces	Piscium	Psc	Fishes

Full Latin name	Genitive form	Symbol	English name
Piscis Austrinus	Piscis Austrini	PsA	Southern Fish
Puppis	Puppis	Pup	Poop; Stern
Pyxis	Pyxidis	Pyx	Mariner's Compass
Reticulum	Reticuli	Ret	Net
Sagitta	Sagittae	Sge	Arrow
Sagittarius	Sagittarii	Sgr	Archer
Scorpius	Scorpii	Sco	Scorpion
Sculptor	Sculptoris	Scl	Sculptor
Scutum	Scuti	Sct	Shield
Serpens	Serpentis	Ser	Serpent
Sextans	Sextantis	Sex	Sextant
Taurus	Tauri	Tau	Bull
Telescopium	Telescopii	Tel	Telescope
Triangulum	Trianguli	Tri	Triangle
Triangulum Australe	Trianguli Australis	TrA	Southern Triangle
Tucana	Tucunae	Tuc	Toucan
Ursa Major	Ursae Majoris	UMa	Great Bear
Ursa Minor	Ursae Minoris	UMi	Little Bear
Vela	Velorum	Vel	Sails
Virgo	Virginis	Vir	Virgin
Volans	Volantis	Vol	Flying Fish
Vulpecula	Vulpeculae	Vul	Fox

Appendix B

Conversion of Normal Calendar
to the Julian Calendar

An indication was given in Part 1, Chapter 1 as to the use of this table. It gives the Julian Date of the first day of each month from 1975 to 1999: for instance, 19 July 1981 corresponds to the Julian Date 2 444 804. Remember that the Julian Day begins at 12 h 00 min UT and not at 00 h 00 min.

	1975	1976	1977	1978	1979
01/01	2 442 413	2 442 778	2 443 144	2 443 509	2 443 874
01/02	2 444	2 809	3 175	3 540	3 905
01/03	2 472	2 838	3 203	3 568	3 933
01/04	2 503	2 869	3 234	3 599	3 964
01/05	2 533	2 899	3 264	3 629	3 994
01/06	2 564	2 930	3 295	3 660	4 025
01/07	2 594	2 960	3 325	3 690	4 055
01/08	2 625	2 991	3 356	3 721	4 086
01/09	2 656	3 022	3 387	3 752	4 117
01/10	2 686	3 052	3 417	3 782	4 147
01/11	2 717	3 083	3 448	3 819	4 178
01/12	2 747	3 119	3 478	3 843	4 208

	1980	1981	1982	1983	1984
01/01	2 444 239	2 444 605	2 444 970	2 445 335	2 445 700
01/02	4 270	4 636	5 001	5 366	5 731
01/03	4 299	4 664	5 029	5 394	5 760
01/04	4 330	4 695	5 060	5 425	5 791
01/05	4 360	4 725	5 090	5 455	5 821
01/06	4 391	4 756	5 121	5 486	5 852
01/07	4 421	4 786	5 151	5 516	5 882
01/08	4 452	4 817	5 182	5 547	5 913
01/09	4 483	4 848	5 213	5 578	5 944
01/10	4 513	4 878	5 243	5 608	5 974
01/11	4 544	4 909	5 274	5 639	6 005
01/12	4 574	4 939	5 304	5 669	6 035

	1985	1986	1987	1988	1989
01/01	2 446 066	2 446 431	2 446 796	2 447 161	2 447 527
01/02	6 097	6 459	6 827	7 192	7 558
01/03	6 125	6 490	6 855	7 221	7 586
01/04	6 156	6 521	6 886	7 252	7 617
01/05	6 186	6 551	6 916	7 282	7 647
01/06	6 217	6 582	6 947	7 319	7 678
01/07	6 247	6 612	6 977	7 343	7 708
01/08	6 278	6 643	7 008	7 374	7 739
01/09	6 309	6 674	7 039	7 405	7 770
01/10	6 339	6 704	7 069	7 435	7 800
01/11	6 370	6 735	7 100	7 466	7 831
01/12	6 400	6 795	7 130	7 496	7 861

	1990	1991	1992	1993	1994
01/01	2 447 892	2 448 257	2 448 622	2 448 988	2 449 353
01/02	7 923	8 288	8 653	9 019	9 384
01/03	7 951	8 316	8 682	9 047	9 412
01/04	7 982	8 347	8 713	9 078	9 443
01/05	8 012	8 377	8 743	9 108	9 473
01/06	8 043	8 408	8 774	9 139	9 504
01/07	8 073	8 438	8 804	9 169	9 534
01/08	8 104	8 469	8 835	9 200	9 565
01/09	8 135	8 500	8 866	9 231	9 596
01/10	8 165	8 530	8 896	9 261	9 626
01/11	8 196	8 561	8 927	9 292	9 657
01/12	8 266	8 591	8 957	9 322	9 687

	1995	1996	1997	1998	1999
01/01	2 449 718	2 450 083	2 450 449	2 450 814	2 451 179
01/02	9 749	0 114	0 480	0 845	1 210
01/03	9 777	0 143	0 508	0 873	1 238
01/04	9 808	0 174	0 539	0 904	1 269
01/05	9 838	0 204	0 569	0 934	1 299
01/06	9 869	0 235	0 600	0 965	1 330
01/07	9 899	0 265	0 630	0 995	1 360
01/08	9 930	0 296	0 661	1 026	1 391
01/09	9 961	0 327	0 692	1 057	1 422
01/10	9 991	0 357	0 722	1 087	1 452
01/11	0 022	0 388	0 753	1 118	1 483
01/12	0 052	0 418	0 783	1 148	1 513

Appendix C

Decimalization of the Day

The decimal parts of the day are given for every 10 min and finer divisions are obtained by interpolation. For instance, if a period of 0.28125 day is required in hours and minutes, the table shows that it must be between 6 h 40 min = 0.27778 day and 6 h 50 min = 0.28472 day. The required period of 0.28125 day is (0.28125 − 0.27778) or 0.00347 greater than 6 h 40 min. Since each minute is 0.000694 day, the extra period is 0.00347/0.000694 min or 5 min. The required period is thus 6 h 45 min.

	0 h 00 min	0 h 10 min	0 h 20 min	0 h 30 min	0 h 40 min	0 h 50 min
0 h	0.00000	0.00694	0.01389	0.02083	0.02778	0.03472
1 h	0.04167	0.04861	0.05556	0.06250	0.06945	0.07639
2 h	0.08334	0.09028	0.09729	0.10417	0.11112	0.11806
3 h	0.12500	0.13194	0.13889	0.14583	0.15278	0.15972
4 h	0.16667	0.17361	0.18056	0.18750	0.19445	0.20139
5 h	0.20834	0.21528	0.22223	0.22917	0.23612	0.24306
6 h	0.25000	0.25694	0.26389	0.27083	0.27778	0.28472
7 h	0.29167	0.29861	0.30556	0.31250	0.31945	0.32639
8 h	0.33334	0.34028	0.34722	0.35416	0.36112	0.36806
9 h	0.37500	0.38194	0.38889	0.39583	0.40278	0.40972
10 h	0.41667	0.42361	0.43056	0.43750	0.44445	0.45139
11 h	0.45834	0.46528	0.47223	0.47917	0.48612	0.49306
12 h	0.50000	0.50694	0.51389	0.52083	0.52778	0.53472
13 h	0.54167	0.54861	0.55556	0.56250	0.56945	0.57639
14 h	0.58334	0.59028	0.59723	0.60417	0.61112	0.61806
15 h	0.62500	0.63194	0.63889	0.64583	0.65278	0.65972
16 h	0.66667	0.67361	0.68056	0.68750	0.69445	0.70139
17 h	0.70834	0.71528	0.72223	0.72917	0.73612	0.74306
18 h	0.75000	0.75694	0.76389	0.77083	0.77778	0.78472
19 h	0.79167	0.79861	0.80556	0.81250	0.81945	0.82639
20 h	0.83334	0.84028	0.84729	0.85416	0.86112	0.86806
21 h	0.87500	0.88194	0.88889	0.89583	0.90278	0.90972
22 h	0.91667	0.92361	0.93055	0.93750	0.94444	0.95139
23 h	0.95833	0.96528	0.97222	0.97017	0.98611	0.99306

Appendix D

The Electromagnetic Spectrum

Stars are now observed in every region of the electromagnetic spectrum from radio waves to gamma rays. It therefore seems useful to indicate the various regions as in the accompanying diagram.

The wavelength is denoted by λ and expressed in metres, and the frequency by ν and expressed in hertz (cycles per second), abbreviated to Hz. The frequency can be found from the wavelength by the simple relationship

$$\nu = c/\lambda$$

where c is the speed of light ($300\,000\,000$ m/s or 3×10^8 m/s).

We give two examples of the use of this equation:

(i) a wavelength of 1 cm ($= 0.01$ m) corresponds to a frequency of

$$\nu = 300\,000\,000/0.01 \text{Hz}$$
$$\nu = 3 \times 10^{10} \text{ Hz}$$

(ii) conversely, a frequency of 3×10^{18} Hz corresponds to a wavelength of

$$\lambda = c/\nu = 3 \times 10^8 / 3 \times 10^{18} \text{m}$$
$$= 10^{-10} \text{ m or 1 Å or 0.1 nm}$$

236

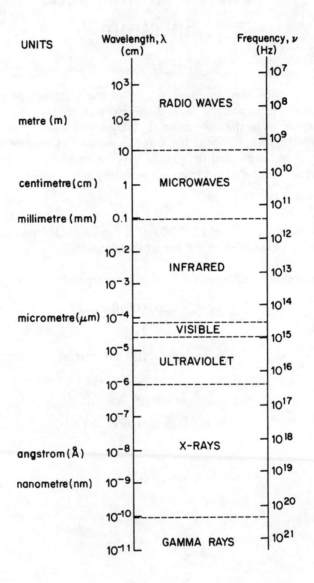

Appendix E

New Classification
of Variable Stars

A new edition of the General Catalogue of Variable Stars was published in June 1985 and included some modifications to the designations given in Table 2 and used throughout this book. The following table gives the new classification and both the old and new designations.

Type	Old designation	New designation
Pulsating variables		
RR Lyrae	RR	RR
with asymmetric light curves	RRab	RRAB
with symmetric light curves	RRc	RRC
with two periods		RR(B)
Cepheids	C	CEP
classical	C	DCEP
classical small amplitude		D CEPS
with two periods		DCEP(B)
W Virginis (Population II)	CW	CW
RV Tauri	RV	RV
Semi-regular	SR	SR
with regular light curves	SRa	SRA
with irregular light curves	SRb	SRB
supergiants	SRc	SRC
yellow (F, G, K spectra)	SRd	SRD
Long-period (Mira)	M	M
Irregular	L	L
giants	Lb	LB
supergiants	Lc	LC
Dwarf Cepheids	RRs	SXPHE
β Canis Majoris	βC	BCEP
δ Scuti	δSc	DSCT
with small amplitude		DSCTC
ZZ Ceti	ZZ	ZZ

Type	Old designation	New designation
Eruptive variables		
Novae	N	N
fast	Na	NA
slow	Nb	NB
recurrent	Nr	NR
Supernovae	SN	SN
Novoids (nova-like variables)	N1	NL
Dwarf novae, U Gem type	UG	UGSS
Z Cam type	Z	UGZ
SU UMa type		UGSU
Nebular variables	In, Is	IN, IS
	UVn	UVN
Red dwarf flare stars	UV	UV

Rotating stars (a category not included in the old classification)

α Canum Venaticorum	CV	ACV
with superposed pulsation		ACVO
Rotating red dwarfs	BY	BY
Giant stars in rapid rotation		FKCOM
Elliptical variables	Ell	ELL

Eclipsing binaries

Algol variables	EA	EA
β Lyrae	EB	EB
W Ursae Majoris	EW	EW

Bibliography

Lack of space prohibits the listing of an extended bibliography and we limit ourselves to an indication of a few general works and the major review journals which deal with variable stars.

GENERAL WORKS

First, there are two older works of historical as well as scientific interest: C. Payne-Gaposchkin and S. Gaposchkin, *Variable Stars*, Harvard Observatory Monograph No. 5, 1938; L. Campbell and L. Jacchia, *The Story of Variable Stars*, Harvard Book on Astronomy, Blackiston, Philadelphia, 1941.

Four more recent works are: J. S. Glasby, *Variable Stars*, Harvard University Press, Cambridge, MA, 1969; W. Strohmeier, *Variable Stars*, Pergamon Press, Oxford, 1972; L. Rosino, *Le Stelle Variabili* (in Italian), published by the Review 'Caelum' of the Societa Astonomica Italiana; C. Hoffmeister, *Veränderliche Sterne* (in German), 1st ed. 1968, 2nd completely revised ed. by G. Richter and W. Wenzel, Johann A. Barth, Leipzig, 1984.

For more detailed points, standard works which are often at an advanced level may be consulted. For example: A. H. Batten, *Binary and Multiple Systems of Stars*, Pergamon Press, Oxford, 1973; J. S. Glasby, *Dwarf Novae*, American Elsevier, New York, 1970; J. S. Glasby, *Variable Nebulae*, Pergamon Press, Oxford, 1974; C. P. Gaposchkin, *Galactic Novae*, Dover, New York, 1964; S. Rosseland, *The Theory of Stellar Pulsations*, Oxford University Press, London, 1949; V. P. Tsesevitch, *Eclipsing Variable Stars*, Wiley, New York, 1973.

Some of the Colloquia of the International Union of Astronomy (IUA) are devoted to variable stars. These are generally published by Reidel, Dordrecht, The Netherlands. The following are of particular interest: No. 4, *Non-Periodic Phenomena in Variable Stars*, 1969; No. 15, *New Directions and New Frontiers in Variable Star Research*, 1971; No. 21, *Variable Stars in Globular Clusters and Related Systems*, 1972; No. 29, *Multiply-Periodic Variable Stars*, 1975; No. 42, *Interaction of Variable Stars with Their Environment*, 1977; No. 46, *Changing Trends in Variable Star Research*, 1978; No. 70, *The Nature of Symbiotic Stars*, 1981.

239

BIBLIOGRAPHICAL WORKS

Three of these are particularly important:

General Catalogue of Variable Stars. The third edition, edited by B. V. Kukarkin in 1969–70 with supplements in 1971, 1974 and 1976, will be replaced by the fourth edition, in the course of publication under the direction of P. N. Kholopov, and published by Sternberg Astronomical Institute, Universitetskii Prospekt 13, 117234 Moscow, USSR. It contains information and a very large bibliography of 28 457 catalogued variable stars.

New Catalogue of Suspected Variable Stars, by P. N. Kholopov *et al.*, 1982, also published by Sternberg Astronomical Institute, Moscow. It contains information about and bibliographical references to 14 810 variable or suspect stars.

Bibliographic Catalogue of Variable Stars, prepared by H. Huth and W. Wenzel and published in 1981 by the Centre de Documentation Stellaire, Strasbourg Observatory, France. It contains a complete bibliography up to 1976 on a large number of variable stars.

JOURNALS

The principal specialized journals are as follows:

Peremennie Zvesdii, publication of the Astronomical Council of the Academy of Sciences of the USSR. Published in Russian with summaries in English by the Sternberg Astronomical Institute, Moscow. It is the only periodical journal devoted entirely to variable stars.

Mitteilungenüber Veränderliche Sterne (in German), published by W. Wenzel, Zentralinstitut für Astrophysik, 64 Sonneberg, GDR. The University of Sonneberg also publishes other original works in *Veröffentlichungen der Sterwarte in Sonneberg*.

Information Bulletin of Variable Stars (*IBVS*), circulars published in English or French and containing original work in a concise form. Published by B. Szeidl and L. Szabados, Konkoly Observatory, P.O. Box 67, H 1525 Budapest, Hungary.

Many articles are published in astronomical journals, particularly in the American journals *The Astrophysical Journal* and *Publications of the Astronomical Society of the Pacific* and in the European Journal *Astronomy and Astrophysics*.

ASSOCIATIONS

Several associations of variable star observers regularly publish their results, as follows:

AFOEV (Association Française d'Observateurs d'Etoiles Variables). Long-period and cataclysmic variables. Publishes a quarterly bulletin. Publisher: E. Schweitzer, 16 rue de Plobsheim, 67100 Strasbourg, France.

AAVSO (American Association of Variable Star Observers). Publishes *The Journal of the AAVSO* and summaries of observations in the *Journal of the Royal Astronomical Society of Canada*, Toronto. Director: J. Mattei, 187 Concord Avenue, Cambridge, MA 02138, USA.

The variable stars section of the British Astronomical Association (BAAVSS). Director J. S. Glasby. Summaries of observations are published in the *Journal of the British Astronomical Association*.

GEOS (Groupe d'Etudes et d'Observations Stellaires). Short-period variables and eclipsing binaries. Results published in *Circulaires, Non-Périodiques*. Publisher: A. Figer, 12 rue Bezout, 75014 Paris, France.

Index of
Astronomical Objects

This index covers all the astronomical objects mentioned in the text and includes in turn:

(i) Galactic variables in alphabetical order of their constellations and, within each constellation, in the order of their official designation (see page 15). The pages in italics are those on which the light curves are to be found.

(ii) Lists of supernovae, clusters, stellar associations, galaxies, diffuse nebulae, planetary nebulae, supernova remnants, radio sources, X-ray sources, active galaxies and quasars and pulsars.

(iii) Lastly, a table giving alternative designations for various objects where there is more than one way of referring to them. These are mainly novae and X-ray sources.

GALACTIC VARIABLES

ALTERNATIVE DESIGNATIONS

Algol	β Per	N. Mon 1939	BT Mon
Maia	20 Tau	N. Oph 1919	V849 Oph
North America	NGC 7000	N. Per 1901	GK Per
Pleiades	M45	N. Pic 1925	RR Pic
Pleione	BU Tau	N. Pup 1942	CP Pup
Praesepe	NGC 2632	N. Sgr 1936	V630 Sgr
Vega	α Lyr	N. Ser 1970	FH Ser
N. Aql 1918	V603 Aql	N. Tau 1927	XX Tau
N. Aql 1927	EL Aql	N. Vul 1968	LV Vul
N. Aql 1936	V368 Aql	SN 1054	CM Tau
N. Aur 1891	T Aur	SN 1572	B Cas
N. Cyg 1920	V476 Cyg	SN 1604	V843 Oph
N. Cyg 1975	V1500 Cyg	SN 166	V598 Cas
N. Del 1967	HR Del	SN 1885a	S And
N. Gem 1903	DM Gem	SN 1895b	Z Cen
N. Gem 1912	DN Gem	SS 433	V1343 Aql
N. Her 1934	DQ Her	Aql X1	V1333 Aql
N. Her 1960	V436 Her	Cen X3	V779 Cen
N. Her 1963	V533 Her	Cir X1	BQ Cir
N. Lac 1910	DI Lac	Cyg X1	V1357 Cyg
N. Lac 1936	CP Lac	Her X1	HZ Her
N. Lac 1950	DK Lac	Sco X1	V818 Sco
N. Mon 1918	GI Mon	Vel X1	GP Vel

General Index

Index of Names